CAREER PLANNING IN ENGINEERING & TECHNOLOGY

CAREER PLANNING IN ENGINEERING & TECHNOLOGY

Naveen Chandra

AKANSHA PUBLISHING HOUSE
NEW DELHI - 110 002 (INDIA)

AKANSHA PUBLISHING HOUSE
4649/21-B, Ansari Road
Daryaganj, New Delhi - 110 002
Phones: 23263193, 9811582579
E-mail: ektabooks@yahoo.com

Career Planning in Engineering & Technology

First Published, 2007

ISBN: 978-81-8370-103-7

PRINTED IN INDIA

Published by M.P. Misra for Akansha Publishing House, New Delhi - 110 002 and Printed at Capital Offset Press, Delhi

Contents

Preface

There is no future in any job; the future lies in the man who holds the job.

—DR. G.W. Crane

The possession of facts is knowledge; the use of them is wisdom; the choice of them, education. Knowledge is not power but riches, and like them, has its value in spending.

—Failure

Educational scenario has changed tremendously over the years. Gone are the days of traditional pattern of education. Now, everyone is conscious of receiving best career-oriented education.

Careers education is defined as "Those formal processes that empower individuals to identify, develop and articulate the skills, qualifications, experiences, attributes and knowledge that will enable them to make an effective transition into their chosen futures, and manage their careers as life long learners, with a realistic and positive attitude".

In a market place where we are aware prospective students and their parents are evaluating the potential outcomes of Higher Education more closely than ever, it is needed to maintain and enhance its position in providing the full range of opportunities for our students. It has become the norm in comparator institutions for careers education to be delivered as part of the curriculum.There are many important skills in engineering – and they're not all technical and scientific. The work needs creative people, organisers and those who are good working with others. Workers that are happy being involved in interpreting drawings and using their practical skills as well as those that enjoy team working, planning and organising people, projects and materials.

Engineering often uses and helps to develop new technology. Engineers ensure that every product that is used is safe, reliable and can be produced and sold at a reasonable cost for the manufacturer and customers, whether it is a telephone, a skateboard or a pair of trainers. An increasing demand for products that are recyclable or not wasteful of the earth's natural energy resources are also issues being tackled by many engineers and the companies that they work for.

This book is designed to be a handbook/guide to all those who aspire to make career in Engineering—an academic discipline, or field of study with bright future prospects. This gives an authoritative account of all major branches of engineering in alphabetic order. Besides the aspirants the information contained herein will enlighten the persons engaged in career guidance and information bureaus.

We acknowledge the assistance, inspiration and guidance received from various quarters while working on this project. The information is gathered from various authoritative sources, we are grateful to them all. Finally, we are thankful to our publisher, for bringing out this book superbly within shortest possible time.

—Editor

Introduction

> Scientists study the world as it is; engineers create the world that has never been.
>
> —*Theodore von Karman*

Engineering is the design, analysis, and/or construction of works for practical purposes. One who practices engineering is called an engineer, and those licensed to do so have formal designations such as Professional Engineer or Chartered Engineer. Engineers use imagination, judgment, and reasoning to apply science, technology, mathematics, and practical experience. The result is the design, production, and operation of useful objects or processes. The broad discipline of engineering encompasses a range of specialized subdisciplines that focus on the issues associated with developing a specific kind of product, or using a specific type of technology.

The history of the concept of "engineering" stems from the earliest times when man began to make clever inventions, such as the pulley, lever, or wheel, etc. The exact etymology of the word engineer, however, is a person occupationally connected with the study, design, and implementation of engines. The word "engine", derives from the Latin *ingenium* (c. 1250), meaning "innate quality, especially mental power, hence a clever invention." Hence, an engineer, essentially, is someone who makes useful or practical inventions.

From another perspective, a now obsolete meaning of engineer, dating from 1325, is "a constructor of military engines". Engineering was originally divided into military engineering, which included construction of fortifications as well as military engines, and civil engineering, non-military construction of such as bridges.

With the rise of engineering as a profession in the nineteenth century the term became more narrowly applied to fields in which mathematics and science were applied to these ends. In some other languages, such as Arabic, the word for "engineering" also means "geometry". Similarly, in addition to military and civil engineering the fields then known as the mechanic arts became incorporated into engineering.

In 1990, with the rise of computer technology, the first search engine was built by computer engineer Alan Emtage.

The crucial and unique task of the engineer is to identify, understand, and interpret the constraints on a design in order to produce a successful result. It is usually not enough to build a technically successful product; it must also meet further requirements. Constraints may include available resources, physical or technical limitations, flexibility for future modifications and additions, and other factors, such as requirements for cost, marketability, producibility, and serviceability. By understanding the constraints, engineers derive specifications for the limits within which a viable object or system may be produced and operated.

Engineers use their knowledge of science, mathematics, and appropriate experience to find suitable solutions to a problem. Creating an appropriate mathematical model of a problem allows them to analyze it (sometimes definitively), and to test potential solutions. Usually multiple reasonable solutions exist, so engineers must evaluate the different design choices on their merits and choose the solution that best meets their requirements. Genrich Altshuller, after gathering statistics on a large number of patents, suggested that compromises are at the heart of "low-level" engineering designs, while at a higher level the best design is one which eliminates the core contradiction causing the problem.

Engineers typically attempt to predict how well their designs will perform to their specifications prior to full-scale production. They use, among other things: prototypes, scale models, simulations, destructive tests, nondestructive tests, and stress tests. Testing ensures that products will perform as expected. Engineers as professionals take seriously their responsibility to produce designs that will perform as expected and will not cause unintended harm to the public at large. Engineers typically include a factor of safety in their designs to reduce the risk of unexpected failure. However, the greater the safety factor, the less efficient the design may be.

Computer Use

As with all modern scientific and technological endeavors, computers and software play an increasingly important role. As well as the typical business application software there are a number of computer aided applications (CAx) specifically for engineering.

One of the most widely used tools in the profession is computer-aided design (CAD) software which enables engineers to create 3D models, 2D drawings, and schematics of their designs. CAD together with Digital mockup (DMU) and CAE software such as finite element method analysis allows engineers to create models of designs that can be analyzed without having to make expensive and time-consuming physical prototypes. These allow products and components to be checked for flaws; assess fit and assembly;

study ergonomics; and to analyze static and dynamic characteristics of systems such as stresses, temperatures, electromagnetic emissions, electrical currents and voltages, digital logic levels, fluid flows, and kinematics. Access and distribution of all this information is generally organized with the use of Product Data Management software.

There are also many tools to support specific engineering tasks such as Computer-aided manufacture (CAM) software to generate CNC machining instructions; Manufacturing Process Management software for production engineering; EDA for printed circuit board (PCB) and circuit schematics for electronic engineers; MRO applications for maintenance management ; and AEC software for civil engineering.

In recent years the use of computer software to aid the development of goods has collectively come to be known as Product Lifecycle Management (PLM).

Engineering in a Social Context

Engineering is a subject that ranges from large collaborations to small individual projects. Almost all engineering projects are beholden to some sort of financing agency: a company, a set of investors, or a government. The few types of engineering that are minimally constrained by such issues are pro bono engineering and open design engineering.

By its very nature engineering is bound up with society and human behaviour. Every product or construction used by modern society will have been influenced by engineering design. Engineering design is a very powerful tool to make changes to environment, society and economies, and its application brings with it a great responsibility, as represented by many of the Engineering Institutions codes of practice and ethics. Whereas medical ethics is a well-established field with considerable consensus, engineering ethics is far less developed, and engineering projects can be subject to considerable controversy. Just a few examples of this from different engineering disciplines are the develoment of nuclear weapons, the Three Gorges Dam, the design and use of Sports Utility Vehicles and the extraction of oil. There is a growing trend amongst western engineering companies to enact serious Corporate and Social Responsibility policies, but many companies do not have these. Engineering is a key driver of human development. Sub-Saharan Africa in particular has a very small engineering capacity and as a result many African nations are unable to implement solutions to problems they face without outside intervention, even if the political and financial obstacles are overcome. The attainment of many of the Millennium Development Goals is primarily an engineering challenge and the achievement of sufficient engineering capacity is a prerequisite to achieving the MDGs.

All overseas development and relief NGOs make considerable use of engineers to apply solutions in disaster and development scenarios. A number of charitable organizations aim to use engineering directly for the good of mankind:

- Engineers Without Borders
- Engineers Against Poverty
- Registered Engineers for Disaster Relief
- Engineers for a Sustainable World

Engineering is a well respected profession. For example, in Canada it ranks as one of the public's most trusted professions. Sometimes engineering has been seen as a somewhat dry, uninteresting field in popular culture, and has also been thought to be the domain of nerds.

Engineers are often respected and ridiculed for their intense beliefs and interests. Perhaps because of their deep understanding of the interconnectedness of many things, engineers such as Governor John H. Sununu, New York City Mayor Michael Bloomberg and Nuclear Physicist Edward Teller, are often driven into politics to "fix things" for the public good.

Occasionally, engineers may be recognized by the "Iron Ring"—a stainless steel or iron ring worn on the little (fourth) finger of the dominant hand. This tradition was originally developed in Canada in the Ritual of the Calling of an Engineer as a symbol of pride and obligation for the engineering profession. Some years later this practice was adopted in the United States. Members of the US Order of the Engineer accept this ring as a pledge to uphold the proud history of engineering. A Professional Engineer's name often has the post-nominal letters PE or P.Eng in North America. In much of Europe a professional engineer is denoted by the letters IR, while in the UK and much of the Commonwealth the term Chartered Engineer applies and is denoted by the letters CEng.

There exists a definitive overlap between the sciences and engineering practice; in engineering, one applies science. Students in mathematics and engineering both begin their education by studying the fundamental principles of mathematics, such as geometry, algebra, and especially, calculus. However, as the students progress, their foci diverge. The mathematics student continues to learn mathematics, consisting of advanced principles and concepts which typically build on those of the past. The engineering student begins to focus primarily on mathematical application. At the end of his education, the engineer will ideally understand most of the science necessary for the solution to physical problems. The scientist will know far more science than the engineer, but with much of this knowledge being too abstract for utility in real world applications.

Engineering is concerned with the design of a solution to a problem. A scientist may ask *why* a phenomenon arises, and proceed to research the answer to the question. Engineers do not deal with scientific phenomena, but may use science in solving a problem, and/or find out *how* to implement that solution. In other words, scientists attempt to *explain* phenomena, whereas engineers use that knowledge to *construct* solutions to problems.

Both areas of endeavor rely on accurate observation of materials and phenomenon. Both use mathematics and classification criteria to analyze and communicate observations. Scientists are expected to interpret their observations and to make expert recommendations for practical action based on those interpretations. Scientists may also have to complete engineering tasks, such as designing experimental apparatus or building prototypes. Conversely, in the process of developing technology engineers sometimes find themselves exploring new phenomena, thus becoming, for the moment, scientists.

However, engineering research has a character different from that of scientific research. First, it often deals with areas in which the basic physics and/or chemistry are well understood, but the problems themselves are too complex to solve in an exact manner. Examples are the use of numerical approximations to the Navier-Stokes equations to describe aerodynamic flow over an aircraft, or the use of Miner's rule to calculate fatigue damage. Second, engineering research employs many semi-empirical methods that are foreign to pure scientific research, one example being the method of parameter variation.

Engineering Education

Education and training in Engineering is being imparted at various levels e.g. certificate, diploma and degree courses. Keeping in view the importance of engineering, many internationally renowned institutes and universities are there imparting quality education to aspirants. In India, IIT's, RECs, and many Technical Universities are there besides some very reputed institutes of Engineering and Technology.

Admission Procedure

Generally, for admission in a bachelor's degree course in any branch of engineering a common entrance examination at state level, IIT's level or individual institution level is held. Every candidate having passed 10+2 exam from CBSE or any other recognized state board of secondary education with Physics, Chemistry and Mathematics as main subjects with certain standard percentage of marks is eligible to appear in the entrance examination.

The date and procedure of Entrance Examination is published in national dailies/newspapers every year. The requisite information can be had from Prospectus of various Institutions, available online too.

Similarly, admission can be sought at any institution abroad, information about which can be had from Internet

After joining an institution the candidate is trained and educated for four years to complete the bachelor's degree course.

On successful completion of the course, the candidates can pursue post graduation and furhter studies in their respective branch of specialisation.

Here in this book, we have given an elaborate introduction all major Engineering branches, in which aspirant can make his career. In strict sense, Engineering education is provided on the following branches: Aerospace engineering, Agricultural engineering, Architectural engineering, Bioengineering, Biochemical engineering, Biomedical engineering, Chemical engineering, Civil engineering, Computer engineering, Construction engineering, Electrical engineering, Electronics engineering, Environmental engineering, Industrial engineering, Materials engineering, Mechanical engineering, Mechatronics engineering, Metallurgical engineering, Mining engineering, Nuclear engineering, Petroleum engineering, Software engineering, Structural engineering and Tissue engineering.

Hopefully, this book will provide ample guidance to different trainers, scholars and aspirants.

—*Ed.*

1

Aerospace Engineering

With tremendous growth of aviation industry, a large number of trained manpower is required. Aerospace engineering provides a bright career for aspirants at global level. Though to touch stars was considered a poetic imagery in olden days, aerospace engineering may equip you to touch stars in reality.

Aerospace engineering is the branch of engineering that concerns aircraft, spacecraft, and related topics. Originally called aeronautical engineering and dealing solely with aircraft, the broader term "aerospace engineering" has replaced the former in most usage, as flight technology advanced to include craft operating outside the Earth's atmosphere. In analogy with "aeronautical engineering", the branch is sometimes referred to as *astronautical engineering*, although this term usually only concerns craft which operate in outer space.

Modern flight vehicles must undergo severe conditions such as differences in atmospheric pressure and temperature, or heavy structural load applied upon vehicle components; numerous matters must be taken into account. Consequently, they are usually products of complex synthesis of various technologies and sciences, including but not limited to aerodynamics, materials science and electronics. The knowledge and the process of combining these various branches of studies are what is collectively known as aerospace engineering. This complex characteristic keeps a single aerospace engineer from involving in the entire task; rather, aerospace engineering is conducted by a team of engineers, each specializing in their own branches of science. The development and manufacturing of a flight vehicle is basically a process to carefully balance and compromise between the abilities, performance, available technology and costs.

The origin of modern-day aerospace engineering can be traced back to the aviation pioneers around the turn of the century from the 19th century to the 20th. Early knowledge of aeronautical engineering was largely empirical with some concepts and skills imported from other branches of engineering, although the early pioneers were not without theoretical background for their creations (fluid dynamics, a key branch of science related to aviation, was present from the century before). Only a decade after the successful flights by the Wright brothers, the 1920s saw extensive development of aeronautical engineering through development of World War I military aircraft. Meanwhile, researches to provide fundamental background science continued by combining theoretical physics with experiments.

Aerospace Engineers

Aerospace engineers design, develop, and test aircraft, spacecraft, and missiles and supervise the production of these products. Those who work with aircraft are called aeronautical engineers, and those working specifically with spacecraft are astronautical engineers. Aerospace engineers develop new technologies for use in aviation, defense systems, and space exploration, often specializing in areas such as structural design, guidance, navigation and control, instrumentation and communication, or production methods.

They also may specialize in a particular type of aerospace product, such as commercial aircraft, military fighter jets, helicopters, spacecraft, or missiles and rockets, and may become experts in aerodynamics, thermodynamics, celestial mechanics, propulsion, acoustics, or guidance and control systems.

Aerospace engineers are expected to have slower-than-average growth in employment over the projection period. Although increases in the number and scope of military aerospace projects likely will generate new jobs, increased efficiency will limit the number of new jobs in the design and production of commercial aircraft. Even with slow growth, the employment outlook for aerospace engineers through 2014 appears favorable: the number of degrees granted in aerospace engineering declined for many years because of a perceived lack of opportunities in this field, and, although this trend is reversing, new graduates continue to be needed to replace aerospace engineers who retire or leave the occupation for other reasons.

Earnings for engineers vary significantly by specialty, industry, and education. Even so, as a group, engineers earn some of the highest average starting salaries among those holding bachelor's degrees.

Elements

Some of the elements of aerospace engineering are:

- Fluid mechanics — the study of fluid flow around objects. Specifically aerodynamics concerning the flow of air over bodies such as wings or through objects such as wind tunnels (see also lift and aeronautics).
- Astrodynamics — the study of orbital mechanics including manipulation, determination, and prediction of orbital elements when given a select

few variables. While few schools in the United States teach this at the undergraduate level, several have graduate programs covering this topic (usually in conjunction with the Physics department of said college or university).

- Dynamics and engineering mechanics — the study of movement, forces, moments in mechanical systems.
- Mathematics — as most subjects within aerospace engineering involve equations and mathematical manipulation and derivations, a solid and comprehensive study of mathematics is required to enable effective learning in the other modules.
- Electrotechnology — the study of electronics within engineering.
- Propulsion — the energy to move a vehicle through the air (or in outer space) is provided by internal combustion engines, jet engines and turbomachinery, or rockets (see also propeller and spacecraft propulsion). A more recent addition to propulsion is ion thrust (or electric) propulsion.
- Control engineering — the study of mathematical modeling of systems and designing them in order that they behave in the desired way. As aircraft flight control systems are becoming increasingly complex, they can be studied as a separate module.
- Aircraft structures — design of the physical configuration of the craft to withstand the forces encountered during flight. Aerospace engineering aims very much at keeping structures lightweight.
- Materials science — related to structures, aerospace engineering also studies the materials of which the aerospace structures are to be built. New materials with very specific properties are invented, or existing ones are modified to improve their performance.

- Aeroelasticity — the interaction of aerodynamic forces and structural flexibility, potentially causing flutter, divergence, etc.
- Avionics — specifically concerning the design and programming of any computer systems on board an aircraft or spacecraft and the simulation of systems. Navigation equipment may be the focus of this study.
- Risk and reliability — the study of risk and reliability assessment techniques and the mathematics involved in the quantitative methods.
- Noise control — the study of the mechanics of sound transfer. Required as noise levels are a massive consideration in the current aerospace industry.
- Flight test — the discipline of designing and executing flight test programs in order to gather and analyze performance and handling qualities data in order to determine if an aircraft meets its design and performance goals and certification requirements.

The basis of most of these elements lies in theoretical mathematics, such as fluid dynamics for aerodynamics or the equations of motion for flight dynamics. However, there is also a large empirical component. Historically, this empirical component was derived from testing of scale models and prototypes, either in wind tunnels or in the free atmosphere. More recently, advances in computing have enabled the use of computational fluid dynamics to simulate the behavior of fluid, reducing time and expense spent on wind-tunnel testing.

Besides, aerospace engineering also addresses the integration of all components that constitute an aerospace vehicle (subsystems including power, communications, thermal control, life support, etc.) and its life cycle (design, temperature, pressure, radiation, velocity, life time), leading to extraordinary challenges and solutions specific to the domain of aerospace systems engineering. It is uncommon for an aerospace engineer to view and comprehend all the components of the involved project.

Aerospace Engineering Degrees

Aerospace (or aeronautical) engineering can be studied at the bachelors, masters, and Ph.D. levels in aerospace engineering departments at many universities, and in mechanical engineering departments at others.

Flight Test

Flight test is a branch of aeronautical engineering that develops and gathers data during flight of an aircraft and then analyses the data to evaluate the flight characteristics of the aircraft and validate its design, including safety aspects. The flight test phase accomplishes two major tasks: 1) finding and fixing any aircraft design problems and then 2) verifying and documenting the aircraft capabilities for government certification or customer acceptance. The flight test phase can range from the test of a single new system for an existing aircraft to the complete development and certification of a new aircraft. Therefore the duration of a flight test program can vary from a few weeks to several years.

There are typically two categories of flight test programs – commercial and military. Commercial flight testing is conducted to certify that the aircraft meets all applicable safety and performance requirements of the government certifying agency. In the US, this is the Federal Aviation Administration (FAA); in Canada, Transport Canada (TC); in the United Kingdom (UK), the Civil Aviation Authority; and in the European Union, the Joint Aviation Authorities (JAA). Since commercial aircraft development is normally funded by the aircraft manufacturer and/or private investors, the certifying agency does not have a stake in the commercial success of the aircraft. These civil agencies are concerned with the aircraft's safety and that the pilot's flight manual accurately reports the aircraft's performance. The market will determine the aircraft's suitability to operators. Normally, the civil certification agency does not get involved in flight testing until the manufacturer has found and fixed any development issues and is ready to seek certification.

Military programs differ from commercial in that the government contracts with the aircraft manufacturer to design and build an aircraft to meet specific mission capabilities. These performance requirements are documented to the manufacturer in the Aircraft Specification and the details of the flight test program (among many other program requirements) are spelled out in the Statement of Work. In this case, the government is the customer and has a direct stake in the aircraft's ability to perform the mission. Since the government is funding the program, it is more involved in the aircraft design and testing from early-on. Often military test pilots and engineers are integrated as part of the manufacturer's flight test team, even before first flight. The final phase of the military aircraft flight test is the Operational Test (OT). OT is conducted by a government-only test team with the dictate to certify that the aircraft is suitable and effective to carry out the intended mission. Flight testing of military aircraft is often conducted at military flight test facilities.

For both commercial and military aircraft, flight test preparation begins well before the aircraft is ready to fly. First, the Flight Test Engineer and the flight test team prepare the test plan(s) to describe the flight maneuvers to be flown, the various aircraft configurations, and the specific data to be taken. A full certification/qualification flight test program for a new aircraft will require testing for many aircraft systems and in-flight regimes; each is typically documented in a separate test plan. During the actual flight testing, similar maneuvers from all test plans are combined and the data collected on the same flights, where practical. This allows the required data to be acquired in the minimum number of flight hours.

Once the flight test data requirements are established, the aircraft is instrumented to record that data for analysis. Typical instrumentation parameters recorded during a flight test are: temperatures, pressures, structural loads, vibration/accelerations, noise levels (interior and exterior), aircraft performance parameters (airspeed, altitude, etc.), aircraft controls positions (stick/yoke position, rudder pedal position, throttle position, etc.), engine performance parameters, and atmospheric conditions. During selected phases of flight test, especially during early development of a new aircraft, many parameters are transmitted to the ground during the flight and monitored by the Flight Test Engineer and test support engineers. This provides for safety monitoring and allows real-time analysis of the data being acquired.

When the aircraft is completely assembled and instrumented, it typically conducts many hours of ground testing before its first/maiden flight. This ground testing will verify basic aircraft systems operations, measure engine performance, evaluate dynamic systems stability, and provide a first look at structural loads. Flight controls will also be checked out. Once all required ground tests are completed, the aircraft is ready for the first flight. First/maiden flight is a major milestone in any aircraft development program and is undertaken with the utmost caution.

There are several aspects to a flight test program: handling qualities, performance, aero-elastic/flutter stability, avionics/systems capabilities, weapons delivery, and structural loads. Handling qualities evaluates the aircraft's controllability and response to pilot inputs throughout the range of flight. Performance testing evaluates aircraft in relation to its projected abilities, such as speed, range, power available, drag, airflow characteristics, and so forth. Aero-elastic stability evaluates the dynamic response of the aircraft controls and structure to aerodynamic (i.e. air-induced) loads. Structural tests measure the stresses on the airframe, dynamic components, and controls to verify structural integrity in all flight regimes. Avionics/ systems testing verifies all electronic systems (navigation, communications, radars, sensors, etc.) perform as designed. Weapons delivery looks at the pilot's ability to acquire the target using on-board systems and accurately deliver the ordnance on target. Weapons delivery testing also evaluates the separation of the ordnance as it leaves the aircraft to ensure there are no safety issues. Other military unique tests are: air-to-air refueling, radar/infrared signature measurement, and aircraft carrier operations. Emergency situations are evaluated as a normal part of all flight test program. Examples are: engine failure during various phases of flight (takeoff, cruise, landing), systems failures, and controls degradation. The overall operations envelope (allowable gross weights, centers-of-gravity, altitude, max/ min airspeeds, maneuvers, etc.) is established and verified during flight testing. Aircraft are always demonstrated to be safe beyond the limits allowed for normal operations in the Flight Manual.

Because the primary goal of a flight test program is to gather accurate engineering data, often on a design that is not fully proven, piloting a flight test aircraft requires a high degree of training and skill, so such programs are typically flown by a specially trained test pilot, and the data is gathered by a flight test engineer, and often visually displayed to the a test pilot and/or flight test engineer using flight test instrumentation.

Major Aerospace Engineering Schools

Argentina

- Universidad Tecnologica Nacional (UTN) - Facultad Regional Haedo Official website

Australia

- Australian Defence Force Academy
- University of Adelaide
- Monash University
- University of New South Wales
- University of Queensland
- Queensland University of Technology
- University of Sydney
- RMIT University

Brazil

- Universidade Federal de Minas Gerais (CEA-UFMG), Belo Horizonte
- Instituto Tecnologico de Aeronautica
- University of Sao Paulo, Sao Carlos

Canada

- Carleton University
- Concordia University (Montreal)
- University of Manitoba (Winnipeg) - an option within Mechanical Engineering, B.Sc.(Eng)
- Ryerson University
- University of Toronto - an option within Engineering

Science, as well as M.A.Sc. and Ph.D. in aerospace engineering

Catalonia

- Universitat Politècnica de Catalunya
- Escola Tècnica Superior d'Enginyeries Industrial i Aeronàutica de Terrassa http://www.etseiat.upc.edu/
- Escola Politècnica Superior de Castelldefels http://www.epsc.upc.edu/

Chile

- Universidad Tecnica Federico•Santa Maria - Academia de Ciencias Aeronauticas

China

- Beijing University of Aeronautics and Astronautics
- Nanjing University of Aeronautics and Astronautics
- Tsinghua University

Chinese Taipei

- Chung Cheng Institute of Technology, National Defense University
- Fneg Chia University
- National Cheng Kung University
- Tamkang University

Egypt

Cairo University - Faculty of Engineering http://www.eng.cu.edu.eg/English.htm

France

- Institut Supérieur de l'Aéronautique et de l'Espace
- École Nationale Supérieure de l'Aéronautique et de l'Espace (SUPAERO)

- École Nationale Supérieure d'Ingénieurs de Constructions Aéronautiques
- École Nationale de l'Aviation Civile
- École Nationale Supérieure de Mécanique et d'Aérotechnique
- Institut de Maintenance Aéronautique de Bordeaux I

Germany

- Technical University of Berlin (TU Berlin)
- Munich University of Technology
- RWTH Aachen University
- Technical University of Brunswick
- University of Stuttgart

Greece

- University of Patras

India

- Punjab Engineering College, Chandigarh
- Indian Institute of Science, Bangalore
- Indian Institute of Technology, Bombay
- Indian Institute of Technology, Kanpur
- Indian Institute of Technology, Kharagpur
- Indian Institute of Technology, Madras
- Madras Institute of Technology, Chennai
- Institute of Aeronautical Engineering, Hyderabad
- Amity School of Engineering, NOIDA
- University of Petroleum & Energy Studies
- Hindustan College of Engineering , Chennai

- Park College of Engineering and Technology, Coimbatore

Indonesia

- Bandung Institute of Technology

Iran

- Amirkabir University of Technology
- K.N.Toosi University of Technology
- Sharif University of Technology
- Malek Ashtar University of Technology

Israel

- Technion-Israel Institute of Technology

Italy

- Second University of Naples http://www.diam.unina2.it/
- University of Bologna
- Politecnico di Milano
- First University of Roma "La Sapienza"
- Politecnico di Torino
- University of Pisa
- University of Padua

Japan

- Daiichi University, College of Technology
- Hiroshima University
- Kanazawa Institute of Technology
- Kyoto University
- Kyushu Institute of Technology
- Kyushu University

- Nagoya University
- National Defense Academy of Japan
- Nihon University
- Nippon Bunri University
- Osaka Prefecture University
- Sojo University
- Teikyo University
- Tohoku University
- Tokai University
- Tokyo Institute of Technology
- Tokyo Metropolitan Institute of Technology
- Tokyo Metropolitan University
- University of Tokyo
- Waseda University

Jordan

- Jordan University for Science and Technology http://www.just.edu.jo/

Lebanon

- University of Balamand http://www.balamand.edu.lb/

Malaysia

- International Islamic University Malaysia
- Universiti Putra Malaysia
- Universiti Sains Malaysia

Netherlands

- Delft University of Technology (ABET)

Philippines

- Airlink International Aviation College

Portugal

- IST - Instituto Superior Tecnico
- Universidade da Beira Interior, Covilha
- Academia da Forca Aerea

Pakistan

- Institute Of Space & Technology, Islamabad (IST)

Singapore

- Nanyang Technological University

South Africa

- The University of the Witwatersrand (WITS)

Spain

- Universidad Politecnica de Madrid
- Escuela Tecnica Superior de Ingenieria Aeronautica http://www.aero.upm.es/
- Escuela Universitaria de Ingenieria Tecnica Aeronautica

Sweden

- Royal Institute of Technology

Turkey

- Middle East Technical University
- Istanbul Technical University

United Kingdom

- University of Bristol
- Coventry University
- Cranfield University
- University of Glamorgan (BEng/BSc)
- University of Glasgow (BSc/BEng/MEng in Aeronautical Engineering or Avionics)

- University of Hertfordshire (BEng/MEng)
- Kingston University
- University of Liverpool
- Loughborough University (BEng/MEng)
- Queens University Belfast (Aeronautical Engineering - MEng/BEng)
- University of Southampton
- University of Sheffield (MEng)
- Imperial College London, University of London (MEng)

United States

- Air Force Institute of Technology (Accreditation Board for Engineering and Technology (ABET))
- University of Alabama in Huntsville (ABET)
- University of Alabama (ABET)
- Arizona State University (ABET)
- University of Arizona (ABET)
- Auburn University (ABET)
- Boston University (ABET)
- California Institute of Technology (ABET)
- California Polytechnic State University, San Luis Obispo (ABET)
- California State Polytechnic University, Pomona (ABET)
- California State University, Long Beach (ABET)
- University of California, Davis (ABET)
- University of California, Irvine (ABET)
- University of California, Los Angeles (ABET)

- University of California, San Diego (ABET)
- Case Western Reserve University (ABET)
- University of Central Florida (ABET)
- University of Cincinnati (ABET)
- Clarkson University (ABET)
- University of Colorado at Boulder (ABET)
- Daniel Webster College
- Embry-Riddle Aeronautical University
- Florida Institute of Technology (ABET)
- University of Florida (ABET)
- Georgia Institute of Technology (ABET)
- University of Illinois at Urbana-Champaign (ABET)
- Illinois Institute of Technology (ABET)
- Iowa State University (ABET)
- University of Kansas (ABET)
- University of Maryland, College Park (ABET)
- Massachusetts Institute of Technology (ABET)
- University of Miami
- University of Michigan (ABET)
- University of Minnesota (ABET)
- Mississippi State University (ABET)
- University of Missouri–Rolla (ABET)
- Naval Postgraduate School (ABET)
- University at Buffalo (ABET)
- North Carolina State University (ABET)
- University of Notre Dame (ABET)
- Ohio State University (ABET)

- Oklahoma State University (ABET)
- University of Oklahoma (ABET)
- Pennsylvania State University (ABET)
- Princeton University (ABET)
- Purdue University (ABET)
- Rensselaer Polytechnic Institute (ABET)
- Rose-Hulman Institute of Technology
- Saint Louis University (ABET)
- San Diego State University (ABET)
- San Jose State University (ABET)
- University of Southern California (ABET)
- Stanford University
- Syracuse University (ABET)
- University of Tennessee at Knoxville (ABET)
- Texas A & M University (ABET)
- University of Texas at Arlington (ABET)
- University of Texas at Austin (ABET)
- Tuskegee University (ABET)
- United States Air Force Academy (ABET)
- United States Naval Academy (ABET)
- Virginia Polytechnic Institute and State University (ABET)
- University of Virginia (ABET)
- University of Washington (ABET)
- West Virginia University (ABET)
- Western Michigan University (ABET)
- Wichita State University (ABET)
- Worcester Polytechnic Institute (ABET)

Major Aerospace Corporations

- Airbus
- Boeing
- Bombardier
- BAE Systems
- EADS (including its Airbus subsidiary)
- Embraer
- United Technologies Corporation
- General Dynamics
- Lockheed Martin
- Northrop Grumman
- Raytheon
- Honeywell

REFERENCES

Career: Aerospace Engineer. *Career Profiles*. The Princeton Review. Retrieved on 2006-10-08.

"Science: Engineering: Aerospace". *Open Site*.

Stanzione, Kaydon Al. (1987). "Engineering". *Encyclopædia Britannica* (15) 18: 560–563.

Van Every, Kermit E.. (1986). "Aeronautical engineering". *Encyclopedia Americana* 1: 226. Grolier Incorporated.

2

Agricultural Engineering

Agricultural engineers develop engineering science and technology in the context of agricultural production and processing and for the management of natural resources.

Agricultural engineers design agricultural machinery and equipment and agricultural structures.Agricultural Engineers may perform tasks as planning, supervising and managing the building of dairy effluent schemes, irrigation, drainage, flood and water control systems, perform environmental impact assessments and interpret research results and implement relevant practices.

Some specialties include power system and machinery design; structures and environmental science; and food and bioprocess engineering. They develop ways to conserve soil and water and to improve the processing of agricultural products.

A large percentage of agricultural engineers work in academia or for government agencies such as the United States Department of Agriculture or state agricultural extension services. Agricultural engineers work in production, sales, management, research and development, or applied science.

Major Agriculture Engineering Colleges in India

- Acharya N G Ranga Agricultural University, College of Agricultural Engineering, Bapatla 522101
- Acharya N G Ranga Agricultural University, Faculty of Agriculture, Rajendranagar, Hyderabad 500030
- Agricultural College and Research Institute, Coimbatore 651003
- Aligarh Muslim University, Aligarh 202002
- Aligarh Muslim University, Centre of Agriculture, Aligarh 202002
- All India Institute of Hygiene and Public Health, Kolkata 700073
- Allahabad Agricultural Institute, P O Agricultural Institute, Allahabad 211007
- Andhra University, Waltair 530003
- Annamalai University, Annamalai Nagar 608002
- Assam Agricultural University, Jorhat 785013
- Assam Agricultural University, Faculty of Agriculture, Jorhat 785013
- Babasahed Bhimrao Ambedkar University, Vidya Vihar, Rae Bareli Road, Lucknow 226025
- Banaras Hindu University, Institute of Agricultural Sciences, Faculty of Agriculture, Varanasi 221005
- Barkatullah University, Hoshangabad Road, Bhopal 462026
- Bharthidasan University, Tiruchirapalli 620024
- Bidhan Chandra Krishi Vishwavidyalaya, Faculty of Agriculture, P O Krishi Vishwavidyalaya, Mohanpur 741252, Distt. Nadia
- Bihar Veterinary College, Patna 800014

- Birsa Agricultural University, Kanke, Ranchi 834006
- Bombay Veterinary Science College, Parel, Mumbai 400012
- Central Agricultural University, Iroisemba, Imphal 795001
- Central Institute of Fisheries Education, (Indian Council of Agricultural Research), Seven Bungalows, J.P. Road, Versova, Mumbai 400061
- Central Marine Fisheries Research Institute, Post Box No. 2704, Dr Salim Ali Road, Ernakulam, Cochin 682031
- Chandra Shekar Azad University of Agriculture and Technology, Kanpur 208002
- Chaudhary Charan Singh Haryana Agricultural University, Hisar 125004
- College of Agricultural Engineering & Technology, Dr. Punjabrao Deshmukh Krishi Vidyapeeth, P O Krishinagar, Akola 444104
- College of Agricultural Engineering & Technology, Chandra Shekhar Azad University of Agriculture and Technology, Kanpur-2
- College of Agricultural Engineering and Technology, Orissa University of Agriculture & Technology, Bhubaneswar 751003
- College of Agricultural Engineering and Technology, Ch. Charan Singh Haryana Agricultural University, Hisar 125004
- College of Agricultural Engineering and Technology, Gujarat Agricultural University, Junagarh 362001
- College of Agricultural Engineering, Jawaharlal Nehru Krishi Vishwavidyalaya, Jabalpur 482004

- College of Agricultural Engineering, Tamil Nadu Agricultural University, Kumulur 621712, Tiruchirapalli
- College of Agricultural Engineering, Punjab Agriculture University, Ludhiana 141004
- College of Agricultural Engineering, Marathwada Krishi Vidyapeeth (MAU), Parbhani 431402
- College of Agricultural Engineering, Mahatma Phule Krishi Vidyapeeth, Rahuri 413722, DIst. Ahmednagar
- College of Agricultural Engineering, Raichur 584101
- College of Agricultural Engineering, Rajendra Agricultural University, Pusa, Samastipur 848125
- College of Agricultural Technology, Marathwada Krishi Vidyapeeth, Parbhani 431402
- College of Agriculture and Regional Research Station, Dharwad 580005
- College of Agriculture, Hebbal 560024
- College of Agriculture, Patnagar 263145
- College of Agriculture, Pune 411005
- College of Dairy Science, Rajasthan Agricultural University, Udaipur 313001
- College of Dairy Technology, Krishakanagar, Labhandi, Raipur 492012
- College of Engineering, Waltair, Visakhapatnam 530003
- College of Forestry, Srinagar
- College of Horticulture and Forestry, Solan 173230
- College of Horticulture, Hyderabad 500030
- College of Horticulture, Rahuri, Maharashtra

- College of Technology & Agricultural Engineering, Rajasthan Agricultural University, Udaipur
- College of Technology, G B Pant University of Agriculture & Technology, Pantnagar 263145
- College of Veterinary & Animal Science, Udgir – 413517 Dist.Latur
- College of Veterinary and Animal Science, Bikaner 334001
- College of Veterinary Research, Pantnagar, Nainital 263145
- College of Veterinary Science and Animal Husbandry, Sardar Krushinagar, Banaskantha 385506
- College of Veterinary Science and Animal Husbandry, Bhubaneswar 751003
- College of Veterinary Science and Animal Husbandry, Jabalpur 482001
- College of Veterinary Science and Animal Husbandry, Mathura 211002
- College of Veterinary Science and Animal Husbandry, Mhow
- College of Veterinary Science and Animal Science, Parbhani 431402
- College of Veterinary Science, Anjore
- College of Veterinary Science, Khanapara, Guwahati 781022
- College of Veterinary Science, Hisar 125004
- College of Veterinary Science, Rajendranagar, Hyderabad 500030
- College of Veterinary Science, Lakhimpur
- College of Veterinary Science, Ludhiana 141004

- College of Veterinary Science, Department of Dairy Science, Tirupati 517502
- College of Veterinary Science, Tirupati 517502
- College of Veterinary Science, Udaigiri
- Dairy Science College, National Dairy Research Institute, Karnal 132001
- Dr Bhim Rao Ambedkar University, Faculty of Agriculture, Senate House, Paliwal Park, Agra 282004
- Dr Panjabrao Deshmukh Krishi Vidyapeeth, Faculty of Agricultural Engineering, PO Krishinagar, Akola 444104
- Dr. Balasaheb Sawant Konkan Krishi Vidyapeeth, Dapoli 415712
- Faculty of Veterinary Science and Animal Husbandry, Kanke, Ranchi 834007
- Faculty of Veterinary Science and Animal Husbandry, Baghi Ali Mardan Khan Nowshera, Srinagar
- Faculty of Veterinary Science and Animal Science, P O Krishi Viswavidyalaya, Nadia 741252
- Forest Research Institute, Dehradun
- G.B. Pant University of Agriculture, Pant Nagar, Nainital 263145
- Goa University, Goa 403005
- Govind Ballabh Pant University of Agriculture & Technology, Faculty of Agriculture, Pantnagar 263145, Distt. Udham Singh Nagar
- Gujarat Agricultural University, Faculty of Agriculture, Sardar Krushinagar 385506, Dist. Banaskantha

- Gujarat College of Veterinary Science and Animal Husbandry, Anand 388001
- Guru Ghasidas University, P.O. Koni, Bilaspur 495009
- Guru Nanak Dev University, Faculty of Agriculture and Forestry, Amritsar 143005
- Himachal Pradesh Krishi Vishwavidyalaya, Faculty of Agriculture, Palampur 176062, Distt. Kangra
- Imphal Fisheries Training College, Barrackpore
- Indian Agriculture Research Institute, New Delhi
- Indira Gandhi Krishi Vishwavidyalaya, Faculty of Agriculture, Krishak Nagar, Raipur 492012
- Jawharlal Nehru Krishi Vishwavidyalaya, Faculty of Agricultural Engineering, Krishi Nagar, Agartala, Jabalpur 482004
- Kelappaji College of Agricultural Engineering and Technology, Kerala Agricultural University, Tavanur 679573, Malapuram Dist.
- Kerala Agricultural University, Vellanikkara 680654
- KNP College of Veterinary Science, Shirval Dist. Satara, Pin. 412 801
- Madras University, Chennai 600005
- Mahatma Gandhi Gramoday Vishwavidyalay, Faculty of Agriculture, Chitrakoot, Distt. Satna 485331
- Mahatma Phule Krishi Vidyapeeth, Faculty of Agricultural Engineering, Rahuri, Dist. Ahmednagar 413722
- Manipur University, Canchipur, Imphal 795003
- Manipur University, Canchipur, Imphal 795003

- Marathwada Krishi Vidyapeeth, Faculty of Agriculture, Krishi Nagar, Parbhani 431402
- Marathwada University, Aurangabad 431004
- Nagaland University, School of Agricultural Science and Rural Development (SASARD) Lumami (HQ) Kohima 797001
- Nagpur Veterinary College, Seminary Hills, High Land drive, Nagpur 440006
- Narendra Deva University of Agriculture & Technology, Faculty of Agriculture, Narendranagar, P O Kumarganj, Faizabad 224229
- North-Eastern Hill University, Faculty of Science, NEHU Campus P.O., Shillong 793022
- Orissa University of Agriculture & Technology, Faculty of Agriculture, Bhubaneswar, Khurda 751003
- Osmania University, Hyderabad 500067
- P.G. Institute Punjabrao Krishi Vidyapeeth, Akola 444001
- Punjab Agriculture University, Faculty of Agriculture, Ludhiana 141004
- Rajasthan Agricultural University, Faculty of Agriculture, Bikaner 334002
- Rajendra Agricultural University, Faculty of Agriculture, P O Pusa, Samastipur 848125
- Rev. Carey Institute of Horticulture, Kolkata 700027
- Sanjay Gandhi Institute of Dairy Science and Tech., Pusa 848125
- Sanjay Gandhi Institute of Dairy Technology, Rajendra Agricultural University, P O Dhelwan, Lohiyanagar, Patna 800020

- Saurashtra University, Kalavad Road, Rajkot 360005
- School of Marine Sciences, Cochin University of Science & Technology, Cochin 682016
- Seth M C College of Dairy Science, Gujarat Agricultural University, Anand Campus, Anand 388110
- Tamil Nadu Agricultural University, Coimbatore 641003
- Tamil Nadu Veterinary and Animal Science University, Vepery High Road, Chennai 600007
- University of Agricultural Sciences Bangalore, Gandhi Krishi Vignana Kendra, Bangalore 560065
- University of Agricultural Sciences Dharwad, Krishnagar, Dharwad 580005
- University of Allahabad, Faculty of Animal Husbandry, Allahabad 211002
- University of Calcutta, Faculty of Agriculture and Veterinary Science, Senate House, 87/1 College Street, Kolkata 700073
- Veterinary College and Research Institute, Namakhal 637002
- Veterinary College, Hebbal, Bangalore 560006
- Veterinary College, Belgocha, Haringhatta
- Visva-Bharati, P O Shantiniketan, Palli-Shiksha Bhawana (Institute of Agriculture), Dist. Birbhum 731235
- West Bengal University of Animal and Fishery Sciences, Faculty of Veterinary and Animal Sciences, 68, Kshudiram Bose Sarani, Kolkata 700037.

3

Architectural Engineering

This world, this universe is the excellent example of nature's architectural engineering.

An architectural engineer applies the skills of many engineering disciplines to the design, construction, operation, maintenance, and renovation of buildings while paying attention to their impacts on the surrounding environment. In countries such as Canada, the UK and Australia, architectural engineering is more commonly known as Building engineering, building systems engineering, or building services engineering. In some languages, such as Korean, "architect" is literally translated as "architectural engineer".

With the establishment of a separate NCEES Professional Engineering registration examination in the 1990s, and first offering in April 2003, architectural engineering is now recognized as a distinct engineering discipline in the United States. But many practicing 'architectural engineers' hold degrees or registration in civil, mechanical, electrical, or another engineering field and become architectural engineers via experience. Conversely, many degree-holding architectural engineers have professional registration in civil or mechanical engineering, for example. The number of architectural engineering degree programs is increasing, but demand far exceeds the availability in the U.S., especially on the East and West Coasts, and in the South.

What differentiates architectural engineering (abbreviated AE, ARE, AREN, or ARCE, for example) from its component engineering disciplines is its appreciation of architecture. Architectural engineers, with their inter-disciplinary background, understand and address architects' goals as they design building systems. Architectural engineers attempt to integrate buildings' HVAC, plumbing, fire protection, electrical, lighting, transportation, and structural systems with the architectural design.

Difference from Architecture

A common confusion is the distinction between architecture and architectural engineering. In essence, architectural engineering is the engineering discipline for the analysis, design, and construction of building systems. Architects are directly responsible for the form and appearance of a building, including the way in which people use and experience the spaces of the building. Architects traditionally act as the leader of the design team, and are thus known as the 'prime professional'. They coordinate the efforts of the various engineering and other design consultants for building projects.

Before about 1975, architectural engineering graduates in the U.S. typically went to work as 'technical architects'. Since that time architectural engineering defined itself as consulting engineers for buildings. Architectural engineers thus concentrate on ensuring that "the buildings work", e.g., that they stand up, that the HVAC systems operate well, that light and electrical power are delivered safely and as needed, and that fire safety is addressed.

By the 1950's, there were approximately 60 architectural engineering degree programs in the U.S. However, as architecture split from engineering — most architecture programs were in engineering schools — many architectural engineering programs lost institutional support. But from a low of eight programs in the early 1980's, and with the redefinition of the discipline as 'engineers for building systems', architectural engineering education is experiencing significant growth. Demand for admission to the programs, and quality of applicants, is very high. The academic honor society for architectural engineering is Phi Alpha Epsilon.

The Architectural Engineer

Architectural engineers' roles can overlap with that of the architect and other project engineers. Like architects, they seek to achieve optimal designs within the overall constraints, except using primarily the tools of engineering rather than architecture. In most parts of the world, architectural engineers are not entitled to practice architecture unless they are also licensed as architects. In some juristictions, registered professional architectural engineers are limited, by virtue of the exams taken, to practicing only one or more of the component areas of building engineering practice such as mechanical (HVAC/plumbing/etc.), electrical, structural, or fire protection.

In recent years there has been increasing emphasis on sustainable and green design, including in engineered building systems. Architectural engineers increasingly seek LEED ((R) USGBC) Accredited Design Professional (LADP) status in addition to their Professional Engineering registration.

Some Potential Career Titles

- ***Consulting Engineer/Design Engineer/Designer.*** An Engineer, usually a Professional Engineer, or if early in his or her career, an Engineer-in-Training that designs and specifies building systems, analyses problems, or optimizes conditions, for example. Typically employed by, or owner of, a consulting engineering firm, but also commonly in an A/E (architectural and engineering services) firm.

- ***Plant/Facilities Engineer.*** The owner's management liaison person interacting with architects, contractors, and engineers in the design and construction of remodeling, additions, and new facilities. Manages and develops such programs within the plant as energy conservation, preventative maintenance, and retrofits.
- ***Sales/Applications Engineer.*** Provides technical advice and application of their representative products to the building industry's architects, engineers, and constructors.
- ***Construction Project Manager.*** Manages the construction of a building project. Responsibilities include the scheduling of labor trades, material, and equipment for the most economical and expeditious construction of buildings.
- ***Construction Estimator.*** The estimator is responsible for the takeoff material, type of labor, and equipment, and calculating the cost for the construction project plus preparing the necessary documents for the estimate.
- ***Structural Engineer.*** Analyzes, calculates, and selects systems and components for various structures. Graduates are employed within the building industry, but also in other structural areas. Graduate study, and additional registration exams in structural engineering are recommended.
- ***Electrical Systems Engineer.*** Designs and specifies electrical power, lighting, and communication systems for buildings. Employed in an electrical consulting design office or in electrical design-construction offices. Other responsibilities may be preparing specifications and cost estimates.
- ***Electrical Utility Engineer.*** Coordinates new

building construction with building owners, design engineers, and contractors, and educates customers on conservation and cost saving opportunities while optimizing the loads on the utility.

- ***Building Inspector.*** Employed by a public agency. Responsible for the public interest to inspect building projects for code compliance.
- ***Fire Safety/Protection Engineer.*** Designs various types of fire protection systems within the building. Systems include sprinkler, chemical suppression, smoke control, and detection devices.
- ***Heating, Ventilating, and Air-Conditioning (HVAC) Engineer.*** Designs the HVAC systems and prepares the specifications.
- ***Plumbing Engineer.*** Designs the potable water, process fluids, and wastewater systems for the buildings.
- ***Professor/Researcher.*** Teaches and performs research and service. Typically requires completion of a Doctor of Philosophy (Ph.D.) in engineering degree.

Major Specializations

- Architecture
- Structural engineering
- Construction engineering
- Construction management
- Project management
- Green building
- Heating, ventilating, and air-conditioning (HVAC)
- Plumbing and piping (hydronics)
- Fire protection engineering

- Acoustics, noise & vibration control
- Building power systems
- Lighting
- Building transportation systems

Typical Coursework in Bachelor Degree Programs

Curricula vary considerably between universities, but the following are often present in ABET-accredited four- or five-year B.S. in architectural engineering degree programs in the U.S.:

- An introduction to architectural engineering
- Two courses in English
- Five courses in mathematics (calculus, differential equations, linear algebra, probability)
- Two courses in physics
- One course in chemistry
- One course in science (e.g., geology, environment, biology)
- Several courses in humanities/social sciences (e.g., architectural history, sociology)
- Two courses in architectural design
- Two courses in building materials and construction

Engineering science courses in:

- Statics and dynamics
- Strength of materials
- Structural analysis
- Basic circuits
- Thermodynamics
- Fluid mechanics

- Heat transfer
- Engineering economics
- Computer programming
- Computer-aided design and drafting (CADD)

Engineering design courses in:

- Structures (e.g., steel, concrete, and/or wood design)
- Construction
- Power and lighting systems
- Plumbing and piping systems
- Heating, ventilating, and air-conditioning (HVAC) systems
- Senior design project
- electives

More Programs Needed

Although the number of architectural engineering degree programs has doubled in the past 20 years, many more programs are needed. In the U.S., programs are plentiful in the Midwest and Plains states but there is unmet demand in much of the Northeast, South, Southwest, and Far West. Institutions that have both engineering and architecture schools already offer most of the required courses; only a modest number of new faculty are needed for the specialty engineering design courses. Universities with only engineering schools will need to hire at least one registered architect as well. Faculty, administrators, or alumni/donors interested in founding architectural engineering degree programs are encouraged to contact the Architectural Engineering Institute, or faculty at current programs, for more information.

High School Preparation

For enhanced success in an architectural engineering bachelor's degree program, at an institution with English instruction, completing at least the follow courses are recommended during a student's high school years:

- One English course each semester
- Mathematics each semester
  - Including geometry, algebra I and II, trigonometry, pre-calculus, and if possible, calculus.
  - Plan on completing calculus as well in college.
- At least one science course per semester, including physics and chemistry
- Typing or wordprocessing
- An art class that is all or part hand-sketching skills
- An introduction to drafting/CADD course, if available
- If available, mechanical, woodworking, and construction courses are also helpful

Major International Architectural Engineering Schools

Africa

- University of Alexandria Department of Architecture, Alexandria, Egypt
- University of Khartoum Department of Architecture, Khartoum, Sudan
- University of Cape Town (UCT) Department of Architecture, Cape Town, South Africa
- University of Benha Department of Architecture, Shoubra, Egypt

- University of Cairo Department of Architecture, Cairo, Egypt

Asia

- School of Architecture - C.E.P.T. University, Ahmedabad, India
- Chu Hai College of Higher Education Department of Architecture, Hong Kong, China
- University of Hong Kong Department of Architecture, Hong Kong, China
- Chinese University of Hong Kong Department of Architecture, Hong Kong, China
- University of Tokyo Department of Architecture, Tokyo, Japan
- School of Planning and Architecture Department of Architecture, New Delhi, India
- Bengal Engineering and Science University Department of Architecture, India

Australia

- Curtin University of Technology, Faculty of Built Environment, Art and Design Perth [www.curtin.edu.au]
- University of Melbourne, Faculty of Architecture, Building and Planning Melbourne [www.abp.unimelb.edu.au]
- RMIT University, Melbourne, Faculty of the Constructed Environment Melbourne [www.fce.rmit.edu.au]
- Deakin University, School of Architecture Geelong
- University of New South Wales, Faculty of the Built Environment Sydney [www.fbe.unsw.edu.au]
- University of Sydney, School of Architecture Sydney

- University of Technology, Sydney, Faculty of Design, Architecture and Building Sydney
- University of Tasmania, Launceston, School of Architecture Launceston [www.utas.edu.au]
- University of Queensland, Brisbane, Faculty of Engineering, Physical Sciences and Architecture Brisbane [www.uq.edu.au]
- Queensland University of Technology, Brisbane, Built Environment and Engineering Brisbane [www.bee.qut.edu.au]

Canada

- University of Waterloo School of Architecture, Waterloo, Canada
- McGill University The School of Architecture, Montreal, Canada
- University of Toronto Faculty of Architecture, Landscape, and Design, Toronto, Canada
- University of British Columbia School of Architecture, Vancouver, Canada
- Carleton University School of Architecture, Ottawa, Canada
- University of Manitoba Faculty of Architecture, Manitoba, Canada
- Dalhousie University (formerly Technical University of Nova Scotia) School of Architecture, Halifax, Canada

Europe

- Architectural Association School of Architecture, London, UK
- Bartlett School of Architecture, Faculty of the Built Environment, University College London, London, UK

- University of Bath Department of Architecture and Civil Engineering, Bath, UK
- Bauhaus, Weimar, Dessau, and Berlin, Germany
- Berlage Institute, Rotterdam, Netherlands
- Canterbury School of Architecture, University College for the Creative Arts, Canterbury, Kent, UK
- Cardiff University Welsh School of Architecture, Cardiff, UK
- Chalmers University of Technology, Department of Architecture, Gothenburg, Sweden
- École des Beaux-Arts, Paris, France
- École Nationale Supérieure d'Architecture de Versailles, Versailles, France
- École Nationale Supérieure d'Architecture de Nantes, Nantes, France
- The University of Edinburgh Architecture, School of Arts Culture & Environment Edinburgh, UK
- EPFL, Lausanne, Switzerland,
- Escola Superior Artística do Porto, Porto, Portugal,
- ETH Zurich, Switzerland,
- Glasgow School of Art, Glasgow, Scotland
- Helsinki University of Technology, Finland.
- Institute for Advanced Architecture of Catalonia (Metapolis), Barcelona, Spain
- Kent School of Architecture, University of Kent, Canterbury, Kent, UK
- Lund University, Department of Architecture, Lund, Sweden
- The Manchester School of Architecture, Manchester, UK

- Politecnico di Milano School of Architecture, Milan, Italy
- Norwegian University of Science and Technology Faculty of Architecture and Fine Art, Trondheim, Norway
- University of Oulu, Finland
- University of Portsmouth, UK
- Royal Institute of Technology, School of Architecture and the Built Environment, Stockholm, Sweden
- RWTH Aachen, Aachen, Germany
- The Scott Sutherland School of Architecture and The Built Environment, Robert Gordon University, Aberdeen, Scotland
- University of Sheffield School of Architecture, Sheffield, UK
- Tampere University of Technology, Finland
- TU Delft Faculty of Architecture, Delft, Netherlands
- University Institute of Architecture, Venice, Italy
- University of Applied Arts Vienna School of Architecture, Vienna, Austria
- University of Architecture, Civil Engineering and Geodesy, Sofia, Bulgaria
- Vienna University of Technology - School of Architecture E253, Vienna, Austria
- Technical University of £ódŸ - Institute of Architecture and Urban Planning, Poland
- The University of Ulster Architecture, School of Architecture Belfast, UK

Mexico

- Universidad Iberoamericana Escuela de Arquitectura, Mexico City, Mexico

- Escuela Superior de Arquitectura, ESArq, Guadalajara

New Zealand

- Unitec New Zealand, Auckland
- University of Auckland, National Institute of Creative Arts and Industries, School of Architecture, Auckland
- Victoria University of Wellington School of Architecture, Wellington

South America

- Universidade Presbiteriana Mackenzie Faculdade de Arquitetura e Urbanismo, Brasil, Brasil
- Universidad de Chile Facultad de Arquitectura y Urbanismo, Santiago, Chile
- Faculdade de Arquitetura da Universidade Federal do Rio Grande do Sul Faculdade de Arquitetura, Porto Alegre, Brasil
- Universidad Nacional de Ingeniería Facultad de Arquitectura, Urbanismo y Artes Lima, Perú
- Universidad de Buenos Aires Facultad de Arquitectura, Diseño y Urbanismo, Buenos Aires, Argentina

4

Biochemical Engineering

Basically, Biochemical Engineering is a branch of chemical engineering that mainly deals with the design and construction of unit processes that involve biological organisms or molecules. Biochemical engineering is often taught as a supplementary option to chemical engineering due to the similarities in both the background subject curriculum and problem-solving techniques used by both professions. Its applications are used in the pharmaceutical, biotechnology, and water treatment industries.

Biochemical Engineering has gained popularity these days. And many aspirants are seeking this as career option.

The Bioreactor

A bioreactor may refer to any device or system that supports a biologically active environment. In one case, a bioreactor is a vessel in which is carried out a chemical process which involves organisms or biochemically active substances derived from such organisms. This process can either be aerobic or anaerobic. These bioreactors are commonly cylindrical, ranging in size from some liter to cube meters, and are often made of stainless steel.

A bioreactor may also refer to a device or system meant to grow cells or tissues in the context of cell culture. These devices are being developed for use in tissue engineering.

On the basis of mode of operation, a bioreactor may be classified as batch, fed batch or continuous (e.g. Continuous stirred-tank reactor model). An example of a bioreactor is the chemostat.

Organism growing in bioreactor may be suspended or immobilized . The simplest, where cells are immobilized, is a Petri dish with agar gel. Large scale immobilized cell bioreactors are: packed bed; fibrous bed; and membrane.

Bioreactor Design

Bioreactor design is quite a complex engineering task. Under optimum conditions the microorganisms or cells are able to perform their desired function with great efficiency. The bioreactor's environmental conditions like gas (i.e., air, oxygen, nitrogen, carbon dioxide) flowrates, temperature, pH and dissolved oxygen levels, and agitation speed/circulation rate need to be closely monitored and controlled.

Most industrial bioreactor manufacturers use vessels, sensors, controllers, and a control system, networked together for their bioreactor system, see programmable logic controller (PLC).

Fouling can harm the overall sterility and efficiency of the bioreactor, especially the heat exchangers. To avoid it the bioreactor must be easily cleanable and must be as smooth as possible (therefore the round shape).

A heat exchanger is needed to maintain the bioprocess at a constant temperature. Biological fermentation is a major source of heat, therefore in most cases bioreactors need water refrigeration. They can be refrigerated with an external jacket or, for very large vessels, with internal coils.

In an aerobic process, optimal oxygen transfer is perhaps

the most difficult task to accomplish. Oxygen is poorly soluble in water -and even less in fermentation broths- and is relatively scarce in air (20.8%). Oxygen transfer is usually helped by agitation, that is also needed to mix nutrients and to keep the fermentation homogeneous. There are however limits to the speed of agitation, due both to high power consumption (which is proportional to the cube of the speed of the electric motor) and the damage to organisms due to excessive tip speed.

Industrial bioreactors usually employ bacteria or other simple organisms that can withstand the forces of agitation. They are also simple to sustain, requiring only simple nutrient solutions and can grow at astounding rates.

In bioreactors where the goal is grow cells or tissues for experimental or therapeutic purposes, the design is significantly different from industrial bioreactors. Many cells and tissues, especially mammalian, must have a surface or other structural support in order to grow, and agitated environments are often destructive to these cell types and tissues. Higher organisms also need more complex growth medium.

5

Biomedical Engineering

Basically, Biomedical engineering (BME) is the application of engineering principles and techniques to the medical field. It combines the design and problem solving expertise of engineering with the medical expertise of physicians to help improve patient health care and the quality of life of healthy individuals. As a relatively new discipline, much of the work in biomedical engineering consists of research and development, covering an array of fields: bioinformatics, medical imaging, image processing, physiological signal processing, biomechanics, biomaterials and bioengineering, systems analysis, 3-D modeling, *etc.* Examples of concrete applications of biomedical engineering are the development and manufacture of biocompatible prostheses, medical devices, diagnostic devices and imaging equipment such as MRIs and EEGs, and pharmaceutical drugs.

Disciplines Within Biomedical Engineering

Clinical Engineering

Clinical engineering is a branch of biomedical engineering for professionals responsible for the management of medical equipment in a hospital. The tasks of a clinical engineer are typically the acquisition and management of medical device inventory, supervising biomedical engineering technicians (BMETs), ensuring that safety and regulatory issues are taken into consideration and serving as a technological consultant for any issues in a hospital where medical devices are concerned. Clinical engineers work closely with the IT department and medical physicists.

A typical biomedical engineering department does the corrective and preventive maintenance on the medical devices used by the hospital, except for those covered by a warantee or maintenance agreement with an external company. All newly acquired equipment is also fully tested. That is, every line of software is executed, or every possible setting is exercised and verified. Most devices are intentionally simplified in some way to make the testing process less expensive, yet accurate. Many biomedical devices need to be sterilized. This creates a unique set of problems, since most sterilization techniques can cause damage to machinery and materials. Most medical devices are either inherently safe, or have added devices and systems so that they can sense their failure and shut down into an unusable, thus very safe state. A typical, basic requirement is that no single failure should cause the therapy to become unsafe at any point during its life-cycle. See safety engineering for a discussion of the procedures used to design safe systems.

Medical Devices

A medical device is intended for use in:

- the diagnosis of disease or other conditions, or
- in the cure, mitigation, treatment, or prevention of disease,
- intended to affect the structure or any function of the body of man or other animals, and which does not achieve any of its primary intended purposes through chemical action and which is not dependent upon being metabolized for the achievement of any of its primary intended purposes.

Some examples include pacemakers, infusion pumps, the heart-lung machine, dialysis machines, artificial organs, implants, artificial limbs, corrective lenses, cochlear implants, ocular prosthetics, facial prosthetics, somato prosthetics, and dental implants.

Medical devices can be regulated and classified (in the US) as shown below:

1. Class I devices present minimal potential for harm to the user and are often simpler in design than Class II or Class III devices. Devices in this category include tongue depressors, bedpans, elastic bandages, examination gloves, and hand-held surgical instruments and other similar types of common equipment.
2. Class II devices are subject to special controls in addition to the general controls of Class I devices. Special controls may include special labeling requirements, mandatory performance standards, and postmarket surveillance. Devices in this class are typically non-invasive and include x-ray machines, PACS, powered wheelchairs, infusion pumps, and surgical drapes.
3. Class III devices require premarket approval, a scientific review to ensure the device's safety and effectiveness, in addition to the general controls of Class I. Examples include replacement heart valves, silicone gel-filled breast implants, implanted cerebellar stimulators, implantable pacemaker pulse generators and endosseous (intra-bone) implants.

Medical Imaging

Imaging technologies are often essential to medical diagnosis, and are typically the most complex equipment found in a hospital including:

- Fluoroscopy
- Magnetic resonance imaging (MRI)
- Nuclear Medicine
- Positron Emission Tomography (PET) PET scansPET-CT scans
- Projection Radiography such as X-rays and CT scans
- Tomography
- Ultrasound
- Electron Microscopy

Tissue Engineering

One of the goals of tissue engineering is to create artificial organs for patients that need organ transplants. Biomedical engineers are currently researching methods of creating such organs. In one case bladders have been grown in lab and transplanted successfully into patients. Bioartificial organs, which utilize both synthetic and biological components, are also a focus area in research, such as with hepatic assist devices that utilize liver cells within an artificial bioreactor construct.

Regulatory Issues

Regulatory issues are never far from the mind of a biomedical engineer. To satisfy safety regulations, most biomedical systems must have documentation to show that they were managed, designed, built, tested, delivered, and used according to a planned, approved process. This is thought to increase the quality and safety of diagnostics and therapies by reducing the likelihood that needed steps can be accidentally omitted again.

In the United States, biomedical engineers may operate under two different regulatory frameworks. Clinical devices and technologies are generally governed by the Food and

Drug Administration (FDA) in a similar fashion to pharmaceuticals. Biomedical engineers may also develop devices and technologies for consumer use, such as physical therapy devices, which may be governed by the Consumer Product Safety Commission.

Other countries typically have their own mechanisms for regulation. In Europe, for example, the actual decision about whether a device is suitable is made by the prescribing doctor, and the regulations are to assure that the device operates as expected. Thus in Europe, the governments license certifying agencies, which are for-profit. Technical committees of leading engineers write recommendations which incorporate public comments and are adopted as regulations by the European Union. These recommendations vary by the type of device, and specify tests for safety and efficacy. Once a prototype has passed the tests at a certification lab, and that model is being constructed under the control of a certified quality system, the device is entitled to bear a CE mark, indicating that the device is believed to be safe and reliable when used as directed.

The different regulatory arrangements sometimes result in technologies being developed first for either the U.S. or in Europe depending on the more favorable form of regulation. Most safety-certification systems give equivalent results when applied diligently. Frequently, once one such system is satisfied, satisfying the other requires only paperwork.

Biomedical Engineering Training

Biomedical engineers combine sound knowledge of engineering and biological science, and therefore tend to have a bachelors of science and advanced degrees from major universities, who are now improving their biomedical engineering curriculum because interest in the field is increasing. Many colleges of engineering now have a biomedical engineering program or department from the undergraduate to the doctoral level. Traditionally, biomedical engineering has been an interdisciplinary field to specialize in after completing an undergraduate degree in a more traditional discipline of engineering or science, the reason for this being the requirement for biomedical engineers to be equally knowledgable in engineering and the biological sciences. However, undergraduate programs of study combining these two fields of knowledge are becoming more widespread. As such, many students also pursue an undergraduate degree in biomedical engineering as a foundation for a continuing education in medical school.

Though the number of biomedical engineers is currently low (under 10,000), the number is expected to rise as modern medicine improves. Currently, according to *U.S. News & World Report*, the program at Johns Hopkins University is ranked first among American universities in the category of bioengineering/biomedical engineering. At the undergraduate level, an increasing number of programs are also becoming recognized by ABET as accredited bioengineering/biomedical engineering programs in the United States. Duke University, ranked second in the U.S. by *U.S. News*, was the first program accredited by the Engineering Council for Profession Development (now ABET) in September of 1972.

Remarkable Facts

- Y.C. Fung considered by many to be the founder of the field of Biomechanics
- Robert Langer - Institute Professor at MIT, runs the largest BME laboratory in the world, pioneer in drug delivery and tissue engineering
- Otto Schmitt (deceased) - biophysicist with significant contributions to BME, working with biomimetics

- Ascher Shapiro (deceased) - Institute Professor at MIT, contributed to the development of the BME field, medical devices (e.g. intra-aortic balloons)
- John G. Webster - a pioneer in the field of instrumentation amplifiers for the recording of electrophysiological signals
- U.A. Whitaker (deceased) - provider of The Whitaker Foundation, which supported research and education in BME by providing over $700 million to various universities, helping to create 30 BME programs and helping finance the construction of 13 buildings

Major Organizations

- Thai Biomedical Engineering Research Society (ThaiBME)
- American College of Clinical Engineering (ACCE)
- Association of Institutions concerned with Medical Engineering (UK)
- Biomed.org
- Biomedical Engineering Meetings Calendar
- Biomedical engineering at the NIH
- Biomedical Engineering website
- Cummings Scientific
- Danish Society for Biomedical Engineering
- Foundation supporting biomedical engineering research
- The Biomedical Engineering Network
- The Biomedical Engineering Society (US)
- The Canadian Medical and Biological Engineering Society
- The Clinical Engineering Society of Ontario (Canada)
- BioMedicalWatch

6

Civil Engineering

Basically, Civil Engineering is a broad field of engineering dealing with the planning, construction, and maintenance of fixed structures, or public works, as they are related to earth, water, or civilization and their processes. Most civil engineering today deals with power plants, bridges, roads, railways, structures, water supply, irrigation, the natural environment, sewer, flood control, transportation and traffic.

Engineering has developed from observations of the ways natural and constructed systems react and from the development of empirical equations that provide bases for design. Civil engineering is the broadest of the engineering fields, partly because it is the oldest of all engineering fields. In fact, engineering was once divided into only two fields - military and civil. Civil engineering was defined to distinguish it from military engineering. Within the US, some federal government funding and organization is still part of the United States Army as the Corps of Engineers. Civil engineering is still an umbrella term, comprised of many related specialties.

Main Sub-disciplines

General civil engineering is concerned with the overall interface of human created fixed projects with the greater world. General civil engineers work closely with surveyors and specialized civil engineers to fit and serve fixed projects within their given site, community and terrain by designing grading, drainage (flood control), pavement, water supply, sewer service, electric and communications supply and land (real property) divisions. General engineers spend much of their time visiting project sites, developing community/neighborhood consensus, and preparing construction plans. General Civil Engineering is also referred to as Site/Civil Engineering; a branch of Civil Engineering that primarily focuses on converting a tract of land from usage to another. Site/Civil Engineers typically apply the principles of Geotechnical Engineering, Structural Engineering, Environmental Engineering, Transportation Engineering and Construction Engineering to residential, commercial, industrial and public works projects of all sizes and levels of construction.

Structural Engineering

Structural engineering is concerned with the structural design and structural analysis of buildings, bridges, and other structures. This involves calculating the stresses and forces that act upon or arise within a structure, and designing the structure to successfully resist those forces and stresses. Resistance to wind and seismic loadings, especially performance near resonant frequencies, which affect the overall stability of a structure are major design concerns. Other factors such as durability and cost are also considered. In addition to design of new buildings, structural engineers may design a seismic retrofit for an existing structure to mitigate undesirable performance during earthquakes.

Fire Protection Engineering

Fire protection engineering, also called 'fire safety engineering' is the practice of application of science and engineering principles and experience to protect people and their environments from the destructive effects of fire. The underlying branches of science are fire science, fire dynamics and chemistry. Fire protection engineers, often civil engineers by training, typically design safeguards that aid in fighting fires, such as alarm, sprinkler, and smoke control systems. They are also involved at the design stage where structural design work is being done, and they provide expert advice on materials choices and in the protection of structural members. They are also employed as fire investigators, including such very large-scale cases as the analysis of the mechanism of the collapse of the World Trade Centre. NASA uses fire protection engineers on its space program to ensure safety.

Geotechnical Engineering

Geotechnical engineering is primarily concerned with the interaction of structures with earth materials, and structures constructed from earth materials. Geotechnical engineers analyse site conditions and design foundations and earthworks. Geotechnical engineers deal with soil properties, soil mechanics, compression and swelling of soils, seepage, slopes, embankments, retaining walls, ground and rock anchors, use of synthetic tensile materials in soil structures, soil-structure interaction, and soil dynamics.

Transportation Engineering

Transportation engineering is concerned with moving people and goods efficiently, safely, and in a manner conducive to a vibrant community. This involves specifying, designing, constructing, and maintaining transportation infrastructure which includes streets, highways, rail systems, airports, ports, and mass transit. It includes areas such as transportation design, transportation planning, traffic engineering, urban engineering, queueing theory, pavement engineering, Intelligent Transportation System (ITS), and infrastructure management.

Environmental engineering deals with the treatment

of chemical, biological, and/or thermal waste, the purification of water and air, and the remediation of contaminated sites, due to prior waste disposal or accidental contamination. Among the topics covered by environmental engineering are pollutant transport, water purification, sewage treatment, and hazardous waste management. Environmental engineers can be involved with pollution reduction, green engineering, and industrial ecology. Environmental engineering also deals with the gathering of information on the environmental consequences of proposed actions and the assessment of effects of proposed actions for the purpose of assisting society and policy makers in the decision making process.

Environmental engineering is the contemporary term for sanitary engineering, though sanitary engineering traditionally had not included much of the hazardous waste management and environmental remediation work covered by the term *environmental engineering*. Some other terms in use are public health engineering and environmental health engineering.

Hydraulic Engineering

Hydraulic engineering is concerned with the flow and conveyance of fluids, principally water. This area of civil engineering is intimately related to the design of pipelines, water distribution systems, drainage facilities (including bridges, dams, channels, culverts, levees, storm sewers), and canals. Hydraulic engineers design these facilities using the concepts of fluid pressure, fluid statics, fluid dynamics, and hydraulics, among others.

Water Resources Engineering

Water resources engineering is concerned with the collection and management of water (as a natural resource). As a discipline it therefore combines hydrology, environmental science, meteorology, geology, conservation, and resource management. This area of civil engineering relates to the prediction and management of both the quality and the quantity of water in both underground (aquifers) and above ground (lakes, rivers, and streams) resources. Water resource engineers analyze and model very small to very large areas of the earth to predict the amount and content of water as it flows into, through, or out of a facility. Although the actual design of the facility may be left to other engineers.

Construction Engineering

Construction engineering involves planning and execution of the designs from transportation, site development, hydraulic, environmental, structural and geotechnical engineers. As construction firms tend to have higher business risk than other types of civil engineering firms, many construction engineers tend to take on a role that is more business-like in nature: drafting and reviewing contracts, evaluating logistical operations, and closely-monitoring prices of necessary supplies.

Materials Science

Civil engineering also includes elements of materials science. Construction materials with broad applications in civil engineering include ceramics such as Portland cement concrete (PCC) and hot mix asphalt concrete, metals such as aluminum and steel, and polymers such as polymethylaccrylate (PMMA) and carbon fibers. Current research in these areas focus around increased strength, durability, workability, and reduced cost.

Surveying

Surveying is the process by which a surveyor measures certain dimensions that generally occur on the surface of the Earth. Modern surveying equipment, such as EDM's,

Total Stations, GPS Surveying and Laser Scanning, allow for remarkably accurate measurement of angular deviation, horizontal, vertical and slope distances. This information is crucial to convert the data into a graphical representation of the Earth's surface, in the form a map. This information is then used by Civil Engineers, Contractors and even realtors to design from, build on, and trade, respectively. Elements of a building or structure must be correctly sized and positioned in relation to each other and to site boundaries and adjacent structures. Civil engineers are trained in the methods of surveying and may seek Professional Land Surveyor status.

Careers

In the United States, there is no one typical career path for Civil Engineers. Most engineering graduates start with jobs of low responsibility, and as they prove their competence, are given more and more responsible tasks, but within each subfield of civil engineering, and even within different segments of the market within each branch, the details of a career path can vary. In some fields and in some firms, entry-level engineers are put to work primarily monitoring construction in the field, serving as the "eyes and ears" of more senior design engineers; while in other areas, entry-level engineers end up performing the more routine tasks of analysis or design. More senior engineers can move into doing more complex analysis or design work, or management of more complex design projects, or management of other engineers, or into specialized consulting, including forensic engineering.

Salaries for Civil Engineers in the United States have typically been lower than those for other fields of engineering, but entry-level salaries are higher than those in most non-engineering fields outside IT.

A popular misconception is that civil engineering is

far from the exciting frontiers in mathematics and computer science. In reality, much of what is now computer science was driven by civil engineering, where structural and network analysis problems required parallel computations and development of advanced algorithms.

Education and Licensure

Before becoming a practicing engineer, civil engineers generally complete tertiary (college or higher) educational requirements, followed by several years of practical experience. Each country, state, or province individually regulates civil engineering practice.

In the U.S., one must become a licensed Professional Engineer to do any civil engineering work affecting the public or to legally represent oneself as a civil engineer. Licensure requirements vary slightly by state, but in all cases entail passing two licensure exams, the Fundamentals of Engineering exam and the Principles and Practice exam (commonly called the PE), and completing a state-mandated number of years of work under the supervision of a licensed Professional Engineer. In addition, an educational requirement must often be met. All states accept a four year Bachelor of Science (BS) or Bachelor of Engineering (BEng) degree in Civil Engineering, from an ABET-accredited program, for their educational requirement. The acceptability of degrees in other fields varies by state; some states allow a person to substitute additional years of supervised work experience for the degree requirement. Although the American Society of Civil Engineers encourages states to raise the educational requirement to a graduate degree, advanced degrees are currently optional for civil engineers in the United States. Graduate study may lead either to a Master of Engineering, which is a Professional Master's degree, or to a Master of Science degree, sometimes followed by a PhD in civil engineering or a sub-discipline.

Texas A&M University in College Station, Texas, has the largest Civil Engineering department in the U.S. with 1,081 undergraduate students enrolled in that major alone (2006). This does not include the 125 Texas A&M undergraduates majoring in ocean engineering, which overlaps heavily with (and descended from) civil engineering.

In Britain current graduates require a MSc, MEng or BEng (Hons) in order to become chartered through the Institution of Civil Engineers. The Institution also allows entrants with substantial experience to apply without this level of formal academic achievement. In practice, most civil engineers in the United Kingdom work without chartered status. Unlike in many other European countries, the term 'Engineer' is not legally protected within the United Kingdom.

In Australia and New Zealand, requirements are typically a four year Bachelor of Engineering (BE) degree which includes a substantial component of work experience (typically around 800 hours (or 20 weeks). Unlike the North American and European degrees, which are highly theoretical, but with a lower emphasis on practicality (whilst studying), the Australian and New Zealand degrees place a higher emphasis on the engineering application. Resultantly when ranked, by theoretical testing and "quality" of education, the Australian and New Zealand universities do not fare as well as their North American and European counterparts. Graduating students, however, are often expected to take up roles equivalent to those that the North American or European engineering would enter into only after several years experience. The opinion as to the benefits of this swing both ways.

In India, education in Civil Engineering is provided at a diploma (a 3 years course) or a Bachelor's degree (a four years course) level. Diploma courses are conducted by

Polytechnics run by various state Boards of Technical Education while BE/BTech courses are available at various Technical Universities, regional college(s) of Engineering, ITIs and other institutes. Master's degree programs and doctoral research programs are also available.

International engineering agreements are designed to allow engineers to practice across international borders. In general, these agreements require both educational competencies and professional experiential competencies.

7

Construction Engineering

In fact, Construction engineering concerns the planning and management of the construction of structures such as highways, bridges, airports, railroads, buildings, dams, and reservoirs. Construction of such projects requires knowledge of engineering and management principles and business procedures, economics, and human behavior. Construction engineers engage in the design of structures temporary, cost estimating, planning and scheduling, materials procurement, selection of equipment, and cost control.

Construction Engineering is differentiated from Construction Management from the standpoint of the use of mathematics, science and engineering to analyze problems and design a construction process. Construction engineers build many of the things that people use every day. Construction engineering involves many aspects of construction including: commercial, residential, bridges, airports, tunnels, and dams. It is an extremely large industry that provides employment and business opportunity to many and continues to grow. Currently there are nearly 6 million people working on construction in the United States.

Career

As also elsewhere, Construction is the largest industry in the United States. It provides jobs to millions ranging in all types of education. Construction makes 14% of the

US Gross National Product. Construction engineering is important to the construction industry because it provides much of the design aspect from the office to the field. Construction engineers follow the plans of architects and sometimes design actual structures. After the structure has been designed the engineers make sure it has been built correctly by testing and overseeing the construction.

Tasks: Construction engineers have a lot of responsibilities in their job. Certain tasks have to be completed everyday in order to get the job done correctly. Analyzing reports is a main part of their job description. They must analyze maps, drawings, blueprints, aerial photography and other topographical information. Construction engineers also have to use computer software to design hydraulic systems and structures while following construction codes. They have to calculate load and grade requirements, liquid flow rates and material stress points to ensure that the structure can withstand stress. Keeping a safe workplace is key to having a successful construction company. It is the construction engineer's job to make sure that everything is conducted correctly. In addition to safety, the construction engineer has to make sure that the site stays clean and sanitary . Surveying the land before construction begins is also a job of the construction engineer. They have to make sure that there are no impediments in the way of where the structure will be built and if there are any they must move them. They also must estimate costs and keep the project under budget. Construction engineers have to test the soils and materials used for adequate strength. Finally, construction engineers have to provide construction information, including repairs and cost changes, to the managers.

Knowledge: Construction engineers build structures that are used by people everyday so they have to be safe and be able to withstand the elements. To complete the job properly construction engineers have to have the knowledge of many different aspects. Those aspects include engineering, technology, design, math, construction, English, customer service, management, transportation, public safety, and computers. They use the engineering, technology, and math aspects to make sure they build the structure to the set standards. They use English, customer service, and management knowledge to deal with the people that could possibly buy the structure. They also use this knowledge to inform the management on how the project is coming along and if any changes are needed.

Skills: Most construction engineers have a love for math and science. In addition to these abilities there are many other skills needed to be a construction engineer. Critical thinking, listening, learning, problem solving, monitoring, and decision making are all very important in construction engineering. Construction engineers have to be able to think about all aspects of a problem and listen to other's ideas so that they can learn everything about a project before it begins. After they have begun a project they must solve the problems that they encounter using math and science. They also have to monitor the workers on the job site for safety and to make sure that the project is on time and done correctly. Whenever a problem occurs it is up to the construction engineer to make the decision on how to fix it.

Abilities: Construction engineers have many different kinds of abilities they use to do their job. They have the abilities to reason, express themselves orally, sense a problem, comprehend (oral and written), order information, speak clearly, and visualize. Construction engineers use these abilities to communicate with other workers and to solve problems. They also have to use there abilities to know what kinds of materials to order and how to get those materials while staying under the budget.

Work Activities: Construction engineers have many activities that they have to do everyday. Those activities include drafting, decision making, computer interaction, communication, documenting, creative thinking, organizing, information collecting, estimating, and analyzing. Construction engineers use drafting to design structures and to show others how to build them. They have to analyze information and make the best decision and solve problems. Computers are an important tool used by construction engineers. They use them to write programs and solve equations. Communication is used everyday to interact with coworkers and supervisors. They have to communicate in person, by telephone, and through e-mail. Documentation is used to record important information that needs to be passed on to management. Most documenting is done in electronic form. Creative thinking is used to come up with new ideas and solve problems. Construction engineers have to be organized to accomplish goals and prioritize jobs. They have to gather information on the task at hand before they can start a project. This will help ensure that the job is completed correctly. In order to keep a project under budget, construction engineers have to estimate costs of materials and workers. Finally, they have to analyze data to find answers to problems they are having on the job site.

Educational Requirements

Construction Engineers are educated to design and build structures that are necessary for everyday life. There are only a handful of schools that offer a major in Construction engineering. Some of the more popular colleges are Purdue University, Iowa State University, Bradley University, California State Polytechnic University-Pomona, and National University. Bradley University offers a Bachelor of Science degree in Construction while Cal Poly Pomona and National University offer a Bachelor of Science degree in Construction Engineering Technology. Cal Poly Pomona, Bradley University and National University are not accredited by ABET to offer degrees in Construction Engineering. However, Cal Poly Pomona is accredited by the Technology Accreditation Committee (T.A.C.) of ABET and their graduates are eligible to sit for the California E.I.T. exam. The University of Southern California offers a Bachelor of Science in Civil Engineering with an option in Construction Engineering. For most Construction engineering jobs a Bachelor of Science is required and some construction experience. In order to work on projects that will be used by the public a construction engineer has to get a professional engineers license. The Fundamentals of Engineering and Principles and Practice exams must be passed, and other requirements met, for a construction engineer to receive a Professional Engineers license.

8

Electrical Engineering

Electrical Engineering is a professional engineering discipline that deals with the study and application of electricity, electronics and electromagnetism. The field first became an identifiable occupation in the late nineteenth century with the commercialization of the electric telegraph and electrical power supply. The field now covers a range of sub-disciplines including those that deal with power, optoelectronics, digital electronics, analog electronics, computer science, artificial intelligence, control systems, electronics, signal processing and telecommunications.

The term *electrical engineering* may or may not encompass *electronic engineering*. Where a distinction is made, electrical engineering is considered to deal with the problems associated with large-scale electrical systems such as power transmission and motor control, whereas electronic engineering deals with the study of small-scale electronic systems including computers and integrated circuits. Another way of looking at the distinction is that electrical engineers are usually concerned with using electricity to transmit energy, while electronics engineers are concerned with using electricity to transmit information. William Gilbert, with his 1600 publication of De Magnete, was the originator of the term "electricity" and many regard him as the father of electrical engineering or of electricity.

Tracing back the history, Electricity has been a subject of scientific interest since at least the 17th century, but it was not until the 19th century that research into the subject started to intensify. Notable developments in this century include the work of Georg Ohm, who in 1827 quantified the relationship between the electric current and potential difference in a conductor, Michael Faraday, the discoverer of electromagnetic induction in 1831, and James Clerk Maxwell, who in 1873 published a unified theory of electricity and magnetism in his treatise on *Electricity and Magnetism.*

During these years, the study of electricity was largely considered to be a subfield of physics. It was not until the late 19th century that universities started to offer degrees in electrical engineering. The Darmstadt University of Technology founded the first chair and the first faculty of electrical engineering worldwide in 1882. In 1883 Darmstadt University of Technology and Cornell University introduced the world's first courses of study in electrical engineering and in 1885 the University College London founded the first chair of electrical engineering in the United Kingdom. The University of Missouri subsequently established the first department of electrical engineering in the United States in 1886.

During this period, the work concerning electrical engineering increased dramatically. In 1882, Edison switched on the world's first large-scale electrical supply network that provided 110 volts direct current to fifty-nine customers in lower Manhattan. In 1887, Nikola Tesla filed a number of patents related to a competing form of power distribution known as alternating current. In the following years a bitter rivalry between Tesla and Edison, known as the "War of Currents", took place over the preferred method of distribution. AC eventually replaced DC for generation and power distribution, enormously extending the range and improving the safety and efficiency of power distribution.

The efforts of the two did much to further electrical engineering—Tesla's work on induction motors and polyphase systems influenced the field for years to come, while Edison's work on telegraphy and his development of the stock ticker proved lucrative for his company, which ultimately became General Electric. However, by the end of the 19th century, other key figures in the progress of electrical engineering were beginning to emerge.

Emergence of Radio and Electronics

During the development of radio, many scientists and inventors contributed to radio technology and electronics. In his classic UHF experiments of 1888, Heinrich Hertz transmitted (via a spark-gap transmitter) and detected radio waves using electrical equipment. In 1895, Nikola Tesla was able to detect signals from the transmissions of his New York lab at West Point (a distance of 80.4 km). In 1897, Karl Ferdinand Braun introduced the cathode ray tube as part of an oscilloscope, a crucial enabling technology for electronic television. John Fleming invented the first radio tube, the diode, in 1904. Two years later, Robert von Lieben and Lee De Forest independently developed the amplifier tube, called the triode. In 1920 Albert Hull developed the magnetron which would eventually lead to the development of the microwave oven in 1946 by Percy Spencer.[10][11] In 1934 the British military began to make strides towards radar (which also uses the magnetron), under the direction of Dr Wimperis culminating in the operation of the first radar station at Bawdsey in August 1936.

In 1941 Konrad Zuse presented the Z3, the world's first fully functional and programmable computer. In 1946 the ENIAC (Electronic Numerical Integrator and Computer) of John Presper Eckert and John Mauchly followed, beginning the computing era. The arithmetic performance of these machines allowed engineers to develop completely new technologies and achieve new objectives, including the Apollo missions and the NASA moon landing.

The invention of the transistor in 1947 by William B. Shockley, John Bardeen and Walter Brattain opened the door for more compact devices and led to the development of the integrated circuit in 1958 by Jack Kilby and independently in 1959 by Robert Noyce.

In 1968 Marcian Hoff invented the first microprocessor at Intel and thus ignited the development of the personal computer. The first realization of the microprocessor was the Intel 4004, a 4-bit processor developed in 1971, but only in 1973 did the Intel 8080, an 8-bit processor, make the building of the first personal computer, the Altair 8800, possible.

Education and Training

Electrical engineers typically possess an academic degree with a major in electrical engineering. The length of study for such a degree is usually four or five years and the completed degree may be designated as a Bachelor of Engineering, Bachelor of Science, Bachelor of Technology or Bachelor of Applied Science depending upon the university.

The degree generally includes units covering physics, mathematics, computer science, project management and specific topics in electrical engineering. Initially such topics cover most, if not all, of the sub-disciplines of electrical engineering. Students then choose to specialize in one or more sub-disciplines towards the end of the degree.

Some electrical engineers also choose to pursue a postgraduate degree such as a Master of Engineering/ Master of Science, a Master of Engineering Management, a Doctor of Philosophy in Engineering or an Engineer's degree. The Master and Engineer's degree may consist of either research, coursework or a mixture of the two. The Doctor of Philosophy consists of a significant research component and is often viewed as the entry point to academia. In the United Kingdom and various other European countries, the Master of Engineering is often considered an undergraduate degree of slightly longer duration than the Bachelor of Engineering.

In most countries, a Bachelor's degree in engineering represents the first step towards professional certification and the degree program itself is certified by a professional body. After completing a certified degree program the engineer must satisfy a range of requirements (including work experience requirements) before being certified. Once certified the engineer is designated the title of Professional Engineer (in the United States, Canada and South Africa), Chartered Engineer (in the United Kingdom, Ireland, India and Zimbabwe), Chartered Professional Engineer (in Australia and New Zealand) or European Engineer (in much of the European Union).

The advantages of certification vary depending upon location. For example, in the United States and Canada "only a licensed engineer may seal engineering work for public and private clients". This requirement is enforced by state and provincial legislation such as Quebec's Engineers Act. In other countries, such as Australia, no such legislation exists. Practically all certifying bodies maintain a code of ethics that they expect all members to abide by or risk expulsion. In this way these organizations play an important role in maintaining ethical standards for the profession. Even in jurisdictions where certification has little or no legal bearing on work, engineers are subject to contract law. In cases where an engineer's work fails he or she may be subject to the tort of negligence and, in extreme cases, the charge of criminal negligence. An engineer's work must also comply with numerous other rules and regulations such as building codes and legislation pertaining to environmental law.

Professional bodies of note for electrical engineers include the Institute of Electrical and Electronics Engineers (IEEE) and the Institution of Electrical Engineers (IEE). The IEEE claims to produce 30 percent of the world's literature in electrical engineering, has over 360,000 members worldwide and holds over 300 conferences annually. The IEE publishes 14 journals, has a worldwide membership of 120,000, and claims to be the largest professional engineering society in Europe. Obsolescence of technical skills is a serious concern for electrical engineers. Membership and participation in technical societies, regular reviews of periodicals in the field and a habit of continued learning are therefore essential to maintaining proficiency.

In countries such as Australia, Canada and the United States electrical engineers make up around 0.25% of the labour force. Outside of these countries, it is difficult to gauge the demographics of the profession due to less meticulous reporting on labour statistics. However, in terms of electrical engineering graduates per-capita, electrical engineering graduates would probably be most numerous in countries such as Taiwan, Japan and South Korea.

Tools and Work

From the Global Positioning System to electric power generation, electrical engineers are responsible for a wide range of technologies. They design, develop, test and supervise the deployment of electrical systems and electronic devices. For example, they may work on the design of telecommunication systems, the operation of electric power stations, the lighting and wiring of buildings, the design of household appliances or the electrical control of industrial machinery. Fundamental to the discipline are the sciences of physics and mathematics as these help to obtain both a qualitative and quantitative description of how such systems will work. Today most engineering work involves the use of computers and it is commonplace to use computer-aided design programs when designing electrical systems. Nevertheless, the ability to sketch ideas is still invaluable for quickly communicating with others.

Although most electrical engineers will understand basic circuit theory (that is the interactions of elements such as resistors, capacitors, diodes, transistors and inductors in a circuit), the theories employed by engineers generally depend upon the work they do. For example, quantum mechanics and solid state physics might be relevant to an engineer working on VLSI (the design of integrated circuits), but are largely irrelevant to engineers working with macroscopic electrical systems. Even circuit theory may not be relevant to a person designing telecommunication systems that use off-the-shelf components. Perhaps the most important technical skills for electrical engineers are reflected in university programs, which emphasize strong numerical skills, computer literacy and the ability to understand the technical language and concepts that relate to electrical engineering.

For most engineers technical work accounts for only a fraction of the work they do. A lot of time is also spent on tasks such as discussing proposals with clients, preparing budgets and determining project schedules.[27] Many senior engineers manage a team of technicians or other engineers and for this reason project management skills are important. Most engineering projects involve some form of documentation and strong written communication skills are therefore very important.

The workplaces of electrical engineers are just as varied as the types of work they do. Electrical engineers may be found in the pristine lab environment of a fabrication plant, the offices of a consulting firm or on site at a mine. During their working life, electrical engineers may find themselves supervising a wide range of individuals including scientists, electricians, computer programmers and other engineers.

Sub-disciplines

Electrical engineering has many sub-disciplines, the most popular of which are listed below. Although there are electrical engineers who focus exclusively on one of these sub-disciplines, many deal with a combination of them. Sometimes certain fields, such as electronic engineering and computer engineering, are considered separate disciplines in their own right.

1. Control

Control engineering focuses on the modelling of a diverse range of dynamic systems and the design of controllers that will cause these systems to behave in the desired manner. To implement such controllers electrical engineers may use electrical circuits, digital signal processors, microcontrollers and PLCs (Programmable Logic Controllers). Control engineering has a wide range of applications from the flight and propulsion systems of commercial airliners to the cruise control present in many modern automobiles. It also plays an important role in industrial automation.

Control engineers often utilize feedback when designing control systems. For example, in an automobile with cruise control the vehicle's speed is continuously monitored and fed back to the system which adjusts the motor's speed accordingly. Where there is regular feedback, control theory can be used to determine how the system responds to such feedback.

2. Power

Power engineering deals with the generation, transmission and distribution of electricity as well as the design of a range of related devices. These include transformers, electric generators, electric motors and power electronics. In many regions of the world, governments maintain an electrical network called a power grid that connects a variety of generators together with users of their energy. Users purchase electrical energy from the grid, avoiding the costly exercise of having to generate their own. Power engineers may work on the design and maintenance of the power grid as well as the power systems that connect to it. Such systems are called *on-grid* power systems and may supply the grid with additional power, draw power from the grid or do both. Power engineers may also work on systems that do not connect to the grid, called *off-grid* power systems, which in some cases are preferable to on-grid systems.

3. Electronics

Electronic Engineering involves the design and testing of electronic circuits that use the properties of components such as resistors, capacitors, inductors, diodes and transistors to achieve a particular functionality. The tuned circuit, which allows the user of a radio to filter out all but a single station, is just one example of such a circuit.

Prior to the World War II, the subject was commonly known as *radio engineering* and basically was restricted to aspects of communications and radar, commercial radio and early television. Later, in post war years, as consumer devices began to be developed, the field grew to include modern television, audio systems, computers and microprocessors. In the mid to late 1950s, the term *radio engineering* gradually gave way to the name *electronic engineering*.

Before the invention of the integrated circuit in 1959, electronic circuits were constructed from discrete components that could be manipulated by humans. These discrete circuits consumed much space and power and were limited in speed, although they are still common in some applications. By contrast, integrated circuits packed a large number—often millions—of tiny electrical components, mainly transistors, into a small chip around the size of a coin. This allowed for the powerful computers and other electronic devices we see today.

4. Microelectronics

Microelectronics engineering deals with the design of very small electronic circuit components for use in an integrated circuit or sometimes for use on their own as a general electronic component. The most common microelectronic components are semiconductor transistors, although all main electronic components (resistors, capacitors, inductors) can be created at a microscopic level.

Microelectronic components are created by chemically fabricating wafers of semiconductors such as silicon (at higher frequencies, gallium arsenide and indium phosphide) to obtain the desired transport of electronic charge and control of current. The field of microelectronics involves a significant amount of chemistry and material science and requires the electronic engineer working in the field to have a very good working knowledge of the effects of quantum mechanics.

5. Signal Processing

Signal processing deals with the analysis and manipulations of signals. Signals can be either analog, in which case the signal varies continuously according to the information, or digital, in which case the signal varies according to a series of discrete values representing the information. For analog signals, signal processing may involve the amplification and filtering of audio signals for audio equipment or the modulation and demodulation of signals for telecommunications. For digital signals, signal processing may involve the compression, error detection and error correction of digitally sampled signals.

6. Telecommunications

Telecommunications engineering focuses on the transmission of information across a channel such as a coax cable, optical fibre or free space. Transmissions across free space require information to be encoded in a carrier wave in order to shift the information to a carrier frequency suitable for transmission, this is known as modulation. Popular analog modulation techniques include amplitude modulation and frequency modulation. The choice of modulation affects the cost and performance of a system and these two factors must be balanced carefully by the engineer.

Once the transmission characteristics of a system are determined, telecommunication engineers design the transmitters and receivers needed for such systems. These two are sometimes combined to form a two-way communication device known as a transceiver. A key consideration in the design of transmitters is their power consumption as this is closely related to their signal strength. If the signal strength of a transmitter is insufficient the signal's information will be corrupted by noise.

7. Instrumentation Engineering

Instrumentation engineering deals with the design of devices to measure physical quantities such as pressure, flow and temperature. The design of such instrumentation requires a good understanding of physics that often extends beyond electromagnetic theory. For example, radar guns use the Doppler effect to measure the speed of oncoming vehicles. Similarly, thermocouples use the Peltier-Seebeck effect to measure the temperature difference between two points.

Often instrumentation is not used by itself, but instead as the sensors of larger electrical systems. For example, a thermocouple might be used to help ensure a furnace's temperature remains constant. For this reason, instrumentation engineering is often viewed as the counterpart of control engineering.

8. Computers

Computers engineering deals with the design of computers and computer systems. This may involve the design of new hardware, the design of PDAs or the use of computers to control an industrial plant. Computer engineers may also work on a system's software. However, the design of complex software systems is often the domain of software engineering, which is usually considered a separate discipline. Desktop computers represent a tiny fraction of the devices a computer engineer might work on, as computer-like architectures are now found in a range of devices including video game consoles and DVD players.

Related Disciplines

Mechatronics is an engineering discipline which deals with the convergence of electrical and mechanical systems. Such combined systems are known as electromechanical systems and have widespread adoption. Examples include automated manufacturing systems, heating, ventilation and air-conditioning systems and various subsystems of aircraft and automobiles.

The term *mechatronics* is typically used to refer to macroscopic systems but futurists have predicted the emergence of very small electromechanical devices. Already such small devices, known as micro electromechanical systems (MEMS), are used in automobiles to tell airbags when to deploy, in digital projectors to create sharper images and in inkjet printers to create nozzles for high-definition printing. In the future it is hoped the devices will help build tiny implantable medical devices and improve optical communication.

Biomedical engineering is another related discipline, concerned with the design of medical equipment. This includes fixed equipment such as ventilators, MRI scanners and electrocardiograph monitors as well as mobile equipment such as cochlear implants, artificial pacemakers and artificial hearts.

9

Electronic Engineering

Electronic engineering is a professional discipline that deals with the behavior and effects of electrons (as in electron tubes and transistors) and with electronic devices, systems, or equipment. The term now also covers a large part of electrical engineering degree courses as studied at most Indian and European universities. Its practitioners are called electronics engineers in Europe. In the Americas and some other parts of the world, the term electrical engineer is used to describe a person doing the same work.

Remarkably, many areas, electronic engineering is considered to be at the same level as electrical engineering, requiring that more general programmes be called electrical and electronic engineering (many UK universities have departments of Electronic and Electrical Engineering). Both define a broad field that encompasses many subfields including those that deal with power, instrumentation engineering, telecommunications, and semiconductor circuit design amongst many others.

The name electrical engineering is still used to cover electronic engineering amongst some of the older (notably American) universities and graduates there are called electrical engineers. In Europe, graduates of electronic engineering are known as *electronics engineers*.

Some people believe the term *electrical engineer* should be reserved for those having specialised in power and heavy current or high voltage engineering, while others believe that power is just one subset of electrical engineering (and indeed the term *power engineering* is used in that industry). Again, in recent years there has been a growth of new separate-entry degree courses such as *information and communication engineering*, often followed by academic departments of similar name.

The modern discipline of electronic engineering was to a large extent born out of radio and television development and from the large amount of Second World War development of defence systems and weapons. In the interwar years, the subject was known as radio engineering and it was only in the late 1950s that the term electronic engineering started to emerge.

In the UK, the subject of electronic engineering became distinct from electrical engineering as a university degree subject around 1960. Students of electronics and related subjects like radio and telecommunications before this time had to enroll in the electrical engineering department of the university as no university had departments of electronics. Electrical engineering was the nearest subject with which electronic engineering could be aligned, although the similarities in subjects covered (except mathematics and electromagnetism) lasted only for the first year of the three-year course.

Early Radio

In 1893, Nikola Tesla made the first public demonstration of radio communication. Addressing the Franklin Institute in Philadelphia and the National Electric Light Association, he described and demonstrated in detail the principles of radio communication. In 1896, Guglielmo Marconi made a wireless radio transmission and went on to develop a practical and widely used radio system. In 1904, John Ambrose Fleming , the first professor of electrical

Engineering at University College London, invented the first radio tube, the diode. One year later, in 1906, Robert von Lieben and Lee De Forest independently developed the amplifier tube, called the triode.

In the interwar years the subject of electronics was dominated by the worldwide interest in *radio* and to some extent telephone and telegraph communications. The terms 'wireless' and 'radio' were then used to refer anything electronic. There were indeed few non-military applications of electronics beyond radio at that time until the advent of television. The subject was not even offered as a separate university degree subject until about 1960.

Prior to the second world war, the subject was commonly known as 'radio engineering' and basically was restricted to aspects of communications and RADAR, commercial radio and early television. At this time, study of radio engineering at universities could only be undertaken as part of a physics degree.

Later, in post war years, as consumer devices began to be developed, the field broadened to include modern TV, audio systems, Hi-Fi and latterly computers and microprocessors. In the mid to late 1950s, the term radio engineering gradually gave way to the name electronic engineering, which then became a stand alone university degree subject, usually taught alongside electrical engineering with which it had become associated due to some similarities. Before the invention of the integrated circuit in 1959, electronic circuits were constructed from discrete components that could be manipulated by humans. These discrete circuits consumed much space and power and were limited in speed although they are still common in some applications. By contrast, integrated circuits packed a large number—often millions—of tiny electrical components, mainly transistors, into a small chip around the size of a coin. This allowed for the powerful computers and other electronic devices we see today.dghj

Tubes or Valves

The Vacuum Tube Detector

The invention of the triode amplifier, generator, and detector enabled audio radio. On Christmas Eve, 1906, Reginald Fessenden (using his heterodyne principle) transmitted the first radio audio broadcast in history from Brant Rock, Massachusetts. Ships at sea heard a broadcast that included Fessenden playing O Holy Night on the violin and reading a passage from the Bible. The first known radio news program was broadcast 31 August 1920 by station 8MK, the unlicensed predecessor of WWJ (AM) in Detroit, Michigan. Regular wireless broadcasts for entertainment commenced in 1922 from the Marconi Research Centre at Writtle near Chelmsford, England. While some early radios used some type of amplification through electric current or battery, through the mid 1920s the most common type of receiver was the crystal set. In the 1920s, amplifying vacuum tubes revolutionized both radio receivers and transmitters.

Phonographs and Radiogrammes

This is the early name for record players or combined radios and record players that had some presence in the inter war years.

Television

In 1931 Manfred von Ardenne introduced the cathode ray tube and thus the electronic television. In the same year, the first successful transatlantic television transmission was made from London to New York. The broadcast television system adopted by the BBC in the UK was the EMI system which was chosen over the inferior mechanical system of John Logie Baird.

Radar and Radio Location

During WW2 many efforts were expended in the electronic location of enemy targets and aircraft. These included radio beam guidance of bombers, electronic counter measures, early radar systems etc. During this time very little if any effort was expended on consumer electronics developments.

Computers

In 1942, Konrad Zuse presented the Z3, the world's first functional computer. In 1946, the ENIAC (Electronic Numerical Integrator and Computer) of John Presper Eckert and John Mauchly followed, beginning the computing era. The arithmetic performance of these machines allowed engineers to develop completely new technologies and achieve new objectives. Early examples include the Apollo missions and the NASA moon landing.

Transistors

The invention of the transistor in 1947 by William B. Shockley, John Bardeen and Walter Brattain opened the door for more compact devices and led to the development of the integrated circuit in 1959 by Jack Kilby.

Microprocessors

In 1968, Marcian Hoff invented the microprocessor at Intel and thus ignited the development of the personal computer. Hoff's invention was part of an order by a Japanese company for a desktop computer, which Hoff wanted to build as cheaply as possible. The first realization of the microprocessor was the Intel 4004, a 4-bit processor, in 1969, but only in 1973 did the Intel 8080, an 8-bit processor, make the building of the first personal computer, the Altair 8800, possible.

Electronics

In the field of electronic engineering, engineers design and test circuits that use the electromagnetic properties of electrical components such as resistors, capacitors, inductors, diodes and transistors to achieve a particular functionality. The tuner circuit, which allows the user of a radio to filter out all but a single station, is just one example of such a circuit.

Electronics is often considered to have begun when Lee De Forest invented the vacuum tube in 1907. Within 10 years, his device was used in radio transmitters and receivers as well as systems for long distance telephone calls. Vacuum tubes remained the preferred amplifying device for 40 years, until researchers working for William Shockley at Bell Labs invented the transistor in 1947. In the following years, transistors made small portable radios, or transistor radios, possible as well as allowing more powerful mainframe computers to be built. Transistors were smaller and required lower voltages than vacuum tubes to work.

Before the invention of the integrated circuit in 1959, electronic circuits were constructed from discrete components that could be manipulated by hand. These non-integrated circuits consumed much space and power, were prone to failure and were limited in speed although they are still common in simple applications. By contrast, integrated circuits packed a large number — often millions — of tiny electrical components, mainly transistors, into a small chip around the size of a coin. This allowed for the powerful computers and other electrical devices we see today.

In designing an integrated circuit, electronics engineers first construct circuit schematics that specify the electrical components and describe the interconnections between them. When completed, VLSI engineers convert the schematics into actual layouts, which map the layers of various conductor and semiconductor materials needed to construct the circuit. The conversion from schematics to layouts can be done by software (see electronic design automation) but very often requires human fine-tuning to decrease space and power consumption. Once the layout is complete, it can be sent to a fabrication plant for manufacturing.

Integrated circuits and other electrical components can then be assembled on printed circuit boards to form more complicated circuits. Today, printed circuit boards are found in most electronic devices including televisions, computers and audio players.

Syllabus

Apart from electromagnetics and network theory, other items in the syllabus are particular to *electronics* engineering course. *Electrical* engineering courses have other specialisms such as machines, power generation and distribution. Note that the following list does not include the large quantity of mathematics (maybe apart from the final year) included in each year's study.

Electromagnetics

Elements of vector calculus: divergence and curl; Gauss' and Stokes' theorems, Maxwell's equations: differential and integral forms. Wave equation, Poynting vector. Plane waves: propagation through various media; reflection and refraction; phase and group velocity; skin depth. Transmission lines: characteristic impedance; impedance transformation; Smith chart; impedance matching; pulse excitation. Waveguides: modes in rectangular waveguides; boundary conditions; cut-off frequencies; dispersion relations. Antennas: Dipole antennas; antenna arrays; radiation pattern; reciprocity theorem, antenna gain.

Network Theory

Network graphs: matrices associated with graphs; incidence, fundamental cut set and fundamental circuit matrices. Solution methods: nodal and mesh analysis. Network theorems: superposition, Thevenin and Norton's maximum power transfer, Wye-Delta transformation. Steady state sinusoidal analysis using phasors. Linear constant coefficient differential equations; time domain analysis of simple RLC circuits, Solution of network equations using Laplace transform: frequency domain analysis of RLC circuits. 2-port network parameters: driving point and transfer functions. State equations for networks.

Electronic Devices and Circuits

Electronic Devices: Energy bands in silicon, intrinsic and extrinsic silicon. Carrier transport in silicon: diffusion current, drift current, mobility, resistivity. Generation and recombination of carriers. p-n junction diode, Zener diode, tunnel diode, BJT, JFET, MOS capacitor, MOSFET, LED, p-I-n and avalanche photo diode, LASERs. Device technology: integrated circuits fabrication process, oxidation, diffusion, ion implantation, photolithography, n-tub, p-tub and twin-tub CMOS process.

Analog Circuits: Equivalent circuits (large and small-signal) of diodes, BJTs, JFETs, and MOSFETs. Simple diode circuits, clipping, clamping, rectifier. Biasing and bias stability of transistor and FET amplifiers. Amplifiers: single-and multi-stage, differential, operational, feedback and power. Analysis of amplifiers; frequency response of amplifiers. Simple op-amp circuits. Filters. Sinusoidal oscillators; criterion for oscillation; single-transistor and op-amp configurations. Function generators and wave-shaping circuits. Power supplies.

Digital circuits: Boolean algebra, minimization of Boolean functions; logic gates digital IC families (DTL,

TTL, ECL, MOS, CMOS). Combinational circuits: arithmetic circuits, code converters, multiplexers and decoders. Sequential circuits: latches and flip-flops, counters and shift-registers. Sample and hold circuits, ADCs, DACs. Semiconductor memories. Microprocessor(8085): architecture, programming, memory and I/O interfacing.

Signals and Systems

Definitions and properties of Laplace transform, continuous-time and discrete-time Fourier series, continuous-time and discrete-time Fourier Transform, z-transform. Sampling theorems. Linear Time-Invariant (LTI) Systems: definitions and properties; casuality, stability, impulse response, convolution, poles and zeros frequency response, group delay, phase delay. Signal transmission through LTI systems. Random signals and noise: probability, random variables, probability density function, autocorrelation, power spectral density.

Control Systems

Basic control system components; block diagrammatic description, reduction of block diagrams. Open loop and closed loop (feedback) systems and stability analysis of these systems. Signal flow graphs and their use in determining transfer functions of systems; transient and steady state analysis of LTI control systems and frequency response. Tools and techniques for LTI control system analysis: root loci, Routh-Hurwitz criterion, Bode and Nyquist plots. Control system compensators: elements of lead and lag compensation, elements of Proportional-Integral-Derivative (PID) control. State variable representation and solution of state equation of LTI control systems.

Communications

Analog communication systems: amplitude and angle

modulation and demodulation systems, spectral analysis of these operations, superheterodyne receivers; elements of hardware, realizations of analog communication systems; signal-to-noise ratio (SNR) calculations for amplitude modulation (AM) and frequency modulation (FM) for low noise conditions. Digital communication systems: pulse code modulation (PCM), differential pulse code modulation (DPCM), delta modulation (DM); digital modulation schemes-amplitude, phase and frequency shift keying schemes (ASK, PSK, FSK), matched filter receivers, bandwidth consideration and probability of error calculations for these schemes. BOLLOX

Education and Training

Electronics engineers typically possess an academic degree with a major in electronic engineering. The length of study for such a degree is usually three or four years and the completed degree may be designated as a Bachelor of Engineering, Bachelor of Science or Bachelor of Applied Science depending upon the university.

The degree generally includes units covering physics, mathematics, project management and specific topics in electrical engineering. Initially such topics cover most, if not all, of the subfields of electronic engineering. Students then choose to specialize in one or more subfields towards the end of the degree.

Some electronics engineers also choose to pursue a postgraduate degree such as a Master of Engineering, a Doctor of Philosophy in Engineering or an Engineer's degree. The Master degree is being introduced in some European and American Universities as a first degree and the differentiation of an engineer with graduate and postgraduate studies is often difficult. In these cases, experience is taken into account. The Master and Engineer's degree may consist of either research, coursework or a mixture of the two. The Doctor of Philosophy consists of a significant research component and is often viewed as the entry point to academia.

In most countries, a Bachelor's degree in engineering represents the first step towards certification and the degree program itself is certified by a professional body. After completing a certified degree program the engineer must satisfy a range of requirements (including work experience requirements) before being certified. Once certified the engineer is designated the title of Professional Engineer (in the United States and Canada), Chartered Engineer (in the United Kingdom, Ireland, India, South Africa and Zimbabwe), Chartered Professional Engineer (in Australia) or European Engineer (in much of the European Union).

Fundamental to the discipline are the sciences of physics and mathematics as these help to obtain both a qualitative and quantitative description of how such systems will work. Today most engineering work involves the use of computers and it is commonplace to use computer-aided design programs when designing electronic systems. Although most electronic engineers will understand basic circuit theory, the theories employed by engineers generally depend upon the work they do. For example, quantum mechanics and solid state physics might be relevant to an engineer working on VLSI but are largely irrelevant to engineers working with macroscopic electrical systems.

Licensure, Certification, and Regulation

Some locations require a license for one to legally be called an electronics engineer, or an engineer in general. For example, in the United States and Canada "only a licensed engineer may seal engineering work for public and private clients". This requirement is enforced by state and provincial legislation such as Quebec's Engineers Act.

In other countries, such as Australia, no such legislation exists. Practically all certifying bodies maintain a code of ethics that they expect all members to abide by or risk expulsion. In this way these organizations play an important role in maintaining ethical standards for the profession. Even in jurisdictions where licenses are not required, engineers are subject to the law. For example, much engineering work is done by contract and is therefore covered by contract law. In cases where an engineer's work fails he or she may be subject to the tort of negligence and, in extreme cases, the charge of criminal negligence. An engineer's work must also comply with numerous other rules and regulations such as building codes and legislation pertaining to environmental law.

In locations where licenses are not required, professional certification may be advantageous.

Professional Bodies

Professional bodies of note for electrical engineers include the Institute of Electrical and Electronics Engineers (IEEE) and the Institution of Electrical Engineers (IEE). The IEEE claims to produce 30 percent of the world's literature in electrical/electronic engineering, has over 360,000 members worldwide and holds over 300 conferences annually. The IEE publishes 14 journals, has a worldwide membership of 120,000, certifies Chartered Engineers in the United Kingdom and claims to be the largest professional engineering society in Europe.

Modern Electronic Engineering

Electronic engineering in Europe is a very broad field that encompasses many subfields including those that deal with, electronic devices and circuit design, control systems, electronics and telecommunications, computer systems, embedded software etc. Many European universities now have departments of Electronics that are completely separate from or have completely replaced their electrical engineering departments.

Summing Up

Electronic engineering involves the design and testing of electronic circuits that use the electronic properties of components such as resistors, capacitors, inductors, diodes and transistors to achieve a particular functionality. Before the invention of the integrated circuit in 1959, electronic circuits were constructed from discrete components. These non-integrated circuits consumed much space and power, and were limited in speed although they are still common in simple applications. By contrast, integrated circuits packed a large number — often millions — of tiny electrical components, mainly transistors, into a small chip around the size of a coin. This allowed for the powerful computers and other electronic devices we see today.

Signal processing deals with the analysis and manipulation of signals. Signals can be either analogue, in which case the signal varies continuously according to the information, or digital, in which case the signal varies according to a series of discrete values representing the information.

For analog signals, signal processing may involve the amplification and filtering of audio signals for audio equipment or the modulation and demodulation of signals for telecommunications. For digital signals, signal processing may involve the compression, error checking and error detection of digital signals.

Telecommunications engineering deals with the transmission of information across a channel such as a co-axial cable, optical fibre or free space.

Transmissions across free space require information to be encoded in a carrier wave in order to shift the information to a carrier frequency suitable for transmission, this is known as modulation. Popular analog modulation techniques include amplitude modulation and frequency modulation. The choice of modulation affects the cost and performance of a system and these two factors must be balanced carefully by the engineer.

Once the transmission characteristics of a system are determined, telecommunication engineers design the transmitters and receivers needed for such systems. These two are sometimes combined to form a two-way communication device known as a transceiver. A key consideration in the design of transmitters is their power consumption as this is closely related to their signal strength. If the signal strength of a transmitter is insufficient the signal's information will be corrupted by noise.

Control engineering has a wide range of applications from the flight and propulsion systems of commercial aeroplanes to the cruise control present in many modern cars. It also plays an important role in industrial automation.

Control engineers often utilize feedback when designing control systems. For example, in a car with cruise control the vehicle's speed is continuously monitored and fed back to the system which adjusts the engine's speed accordingly. Where there is regular feedback, control theory can be used to determine how the system responds to such feedback. *Read more...*

Instrumentation engineering deals with the design of devices to measure physical quantities such as pressure, flow and temperature. These devices are known as instrumentation.

The design of such instrumentation requires a good understanding of physics that often extends beyond electromagnetic theory. For example, radar guns use the Doppler effect to measure the speed of oncoming vehicles. Similarly, thermocouples use the Peltier-Seebeck effect to measure the temperature difference between two points.

Often instrumentation is not used by itself, but instead as the sensors of larger electrical systems. For example, a thermocouple might be used to help ensure a furnace's temperature remains constant. For this reason, instrumentation engineering is often viewed as the counterpart of control engineering. *Read more...*

Computer engineering deals with the design of computers and computer systems. This may involve the design of new hardware, the design of PDAs or the use of computers to control an industrial plant. Computer engineers may also work on a system's software. However, the design of complex software systems is often the domain of software engineering, which is usually considered a separate discipline.

Desktop computers represent a tiny fraction of the devices a computer engineer might work on, as computer-like architectures are now found in a range of devices including video game consoles and DVD players. *Read more...*

Project Engineering

For most engineers not involved at the cutting edge of system design and development, technical work accounts for only a fraction of the work they do. A lot of time is also spent on tasks such as discussing proposals with clients, preparing budgets and determining project schedules. Many senior engineers manage a team of technicians or other engineers and for this reason project management skills are important. Most engineering projects involve some form of documentation and strong written communication skills are therefore very important.

The workplaces of electronics engineers are just as varied as the types of work they do. Electronics engineers may be found in the pristine laboratory environment of a fabrication plant, the offices of a consulting firm or in a research laboratory. During their working life, electronics engineers may find themselves supervising a wide range of individuals including scientists, electricians, computer programmers and other engineers.

Obsolescence of technical skills is a serious concern for electronics engineers. Membership and participation in technical societies, regular reviews of periodicals in the field and a habit of continued learning are therefore essential to maintaining proficiency. And these are mostly used in the field of consumer electronics products.

Universities with Separate Electronics (Electronic Engineering) Departments

- University of Regina, CA
- Queens University Belfast, QUB
- UCNW Bangor
- Saint Petersburg Electrotechnical University, Russia
- UCL, UK
- York, UK
- Queen Mary, UK
- Surrey, UK
- Portsmouth, UK
- Dundee, UK
- Kent, UK

- Solent, UK
- Southampton, UK
- Electronica, Romania — warning: Java applet
- University of Canterbury, New Zealand
- Istanbul Technical University, Turkey
- Simón Bolívar University, Venezuela
- National University, Colombia
- Hong Kong Polytechnic University, Hong Kong
- ENSEIRB-Universite de Bordeaux-, France
- Uludag University, Turkey
- Universidad de Buenos Aires - Facultad de Ingeniería, Argentina
- UTN, Argentina
- UTFSM, Chile
- Technical University of Crete, Greece

10

Environmental Engineering

Basically, Environmental engineering is the application of science and engineering principles to improve the environment (air, water, and/or land resources), to provide healthful water, air, and land for human habitation and for other organisms, and to remediate polluted sites. Negative environmental effects can be decreased and controlled through public education, conservation, regulations, and the application of good engineering practices. In the U.S., minimum education requirements for environmental engineers typically include a Bachelor's Degree in environmental (or civil) engineering from an accredited college.

Ever since people first recognized that their health and well-being were related to the quality of their environment, they have applied thoughtful principles to attempt to improve the quality of their environment. The Romans constructed aqueducts to prevent drought and to create a clean, healthful water supply for the metropolis of Rome. In the 15th century, Bavaria created laws restricting the development and degradation of alpine country that constituted the region's water supply.

Modern environmental engineering began in London in the mid-19th century when it was realized that proper sewerage could reduce the incidence of waterborne diseases such as cholera. The introduction of drinking water treatment and sewage treatment in industrialized countries reduced waterborne diseases from leading causes of death to rarities.

In many cases, as societies grew, actions that were intended to achieve benefits for those societies had longer-term impacts which reduced other environmental qualities. One example is the widespread application of DDT to control agricultural pests in the years following World War II. While the agricultural benefits were outstanding and crop yields increased dramatically, thus reducing world hunger substantially, and malaria was controlled better than it ever had been, numerous species were brought to the verge of extinction due to the impact of the DDT on their reproductive cycles. The story of DDT as vividly told in Rachel Carson's "Silent Spring" is considered to be the birth of the modern environmental movement and the development of the modern field of "environmental engineering."

Conservation movements and laws restricting public actions that would harm the environment have been developed by various societies for millennia. Notable examples are the laws decreeing the construction of sewers in London and Paris in the 19th century and the creation of the U.S. national park system in the early 20th century.

Briefly speaking, the main task of environmental engineering is to protect (from further degradation), preserve (the present condition), and enhance (the environment).

Scope

"Pollutants" may be chemical, biological, thermal, radioactive, or even mechanical. Environmental engineering emphasizes several areas: process engineering, environmental chemistry, water and wastewater treatment (sanitary engineering), waste reduction/management, and pollution prevention/cleanup. Environmental engineering is a synthesis of various disciplines, incorporating elements from the following:

- Civil engineering
- Chemical engineering
- Public health
- Mechanical engineering
- Chemistry
- Biology
- Geology
- Ecology

Environmental engineering is the application of science and engineering principles to the environment. Some consider 'Environmental Engineering' to include the development of sustainable processes. There are several divisions of the field of environmental engineering.

Environmental Impact Assessment and Mitigation

It is a decision making tool. In this division, engineers and scientists assess the impacts of a proposed project on environmental conditions. They apply scientific and engineering principles to evaluate if there are likely to be any adverse impacts to water quality, air quality, habitat quality, flora and fauna, agricultural capacity, traffic impacts, social impacts, ecological impacts, noise impacts, visual(landscapc) impacts, etc. If impacts are expected, they then develop mitigation measures to limit or prevent such impacts. An example of a mitigation measure would be the creation of wetlands in a nearby location to mitigate the filling in of wetlands necessary for a road development if it is not possible to reroute the road.

Water Supply and Treatment

Engineers and scientists work to secure water supplies for potable and agricultural use. They evaluate the water balance within a watershed and determine the available water supply, the water needed for various needs in that watershed, the seasonal cycles of water movement through the watershed and they develop systems to store, treat, and convey water for various uses. Water is treated to achieve water quality objectives for the end uses. In the case of potable water supply, water is treated to minimize risk of infectious disease transmittal, risk of non-infectious illness, and create a palatable water flavor. Water distribution systems are designed and built to provide adequate water pressure and flow rates to meet various end-user needs such as domestic use, fire suppression, and irrigation.

Wastewater Conveyance and Treatment

Most urban and many rural areas no longer discharge human waste directly to the land through outhouse, septic, and/or honey bucket systems, but rather deposit such waste into water and convey it from households via sewer systems. Engineers and scientists develop collection and treatment systems to carry this waste material away from where people live and produce the waste and discharge it into the environment. In developed countries, substantial resources are applied to the treatment and detoxification of this waste before it is discharged into a river, lake, or ocean system. Developing nations are striving to obtain the resources to develop such systems so that they can improve water quality in their surface waters and reduce the risk of water-borne infectious disease.

There are numerous wastewater treatment technologies. A wastewater treatment train can consist of a primary clarifier system to remove solid and floating materials, a secondary treatment system consisting of an aeration basin followed by flocculation and sedimentation or an activated sludge system and a secondary clarifier, a tertiary biological nitrogen removal system, and a final disinfection process. The aeration basin/activated sludge system removes organic material by growing bacteria (activated sludge). The secondary clarifier removes the activated sludge from the water. The tertiary system, although not always included due to costs, is becoming more prevalent to remove nitrogen and phosphorus and to disinfect the water before discharge to a surface water stream or ocean outfall.

Air Quality Management

Engineers apply scientific and engineering principles to the design of manufacturing and combustion processes to reduce air emissions to acceptable levels. Scrubbers, precipitators, after-burners, and other devices are utilized to remove particulate matter, nitrogen oxides, sulfur oxides, and reactive organic gases from vapors prior to allowing their emission to the atmosphere. This field is beginning to overlap with energy efficiency and the desire to decrease carbon dioxide and other greenhouse gas emissions from combustion processes. Scientists develop dispersion models to evaluate the concentration of a pollutant at a receptor source or the impact on overall air quality and smog production from vehicle and flue gas stack emissions.

Other Applications

- Contaminated land management and site remediation
- Risk assessment
- Environmental policy and regulation development
- Solid waste management
- Hazardous waste management
- Environmental health and safety

- Natural resource management
- Noise pollution
- Geographic information system (GIS)

Major Environmental Engineering Colleges
Specialisations B.Sc. (Envtal.Sc)

- Berhampur University, Berhampur - 760 007, Orissa
- Bharathiar University, Coimbatore - 641 046
- Manipur University, Imphal - 795 003
- Patna University, Patna - 800 005
- South Gujarat University, Surat - 395 007
- University of Delhi, Delhi - 110 007
- University of Mysore, Mysore - 570 005
- University of Pune, Pune - 411 007

B.E.With Specialisation in Environmental Studies

- Andhra University, College of Engineering, Vishakhapatnam - 530 003
- Delhi College of Engineering, New Delhi - 110 042
- Gujarat University, L.D. College of Engineering, Ahmedabad - 380 015
- Kolhapur Institute of Technology's College of Engineering, Kolhapur - 416 234
- PES College of Engineering, Mandya-571 401
- Vidya Vardhaka College of Engineering, Gokulam III Stage, Mysore - 570 002
- Visveswariah Technological University, AIT, Chikmagalur - 577 102

PG Diploma programmes in Environment Management and Environmental Science

- Indian Institute of Environment Management (SIES), Sri Chandrasekarendra Saraswathy Vidyapuram, Plot 1E, Sector V, Nerul, Navi Mumbai - 400 706
- SNDT Women's University, 1, Nathibai Thackersey Road, New Marine Lines, Mumbai - 20
- University of Chennai, Chennai - 600 005
- Vikram University, Ujjain, Madhya Pradesh - 456 010

Institutions offering Degree and PG course in Environmental Studies

- Andhra University, Waltair, Andhra Pradesh
- Annamalai University, Annamalai Nagar, Tamil Nadu
- Avadh University, Faizabad, Uttar Pradesh
- Awadesh Pratap Singh University, Rewa, MP
- Bharathidasan University, Palkalai Perun, Timchinapath, Tamil Nadu
- Centre for Ecological Science, Indian Institute of Science, Bangalore - 560 012
- Centre for Environment Education, Thaltes Tekar, Ahmedabad, Gujarat
- Centre for Environmental Studies, Indian Institute of Technology, Powai, Mumbai - 400 076
- Department of Environmental Biology, University of Delhi, South Campus, New Delhi - 110 021
- Department of Environmental Studies, Government Engineering College, Trichur, Kerala
- Department of Environmental Studies, Guwahati University, Guwahati

- Dr.Harisingh Gaur Vishwavidyalaya, Shivaji University, Vidyanagar, Kolhapur, Maharashtra
- GB Pant University of Agriculture and Technology, Pantnagar, Uttar Pradesh
- Indian Institute of Ecology and Environment, Maidan Garhi Road, New Delhi - 110 030
- Indian School of Mines, Dhanbad - 826 004, Bihar
- Institute of environment Education, Bharathi Vidyapeeth, Pune - 411 043
- Institute of Social Welfare and Business Management, Management House, College Square West, Calcutta - 73
- Jadavpur University, Jadavpur, Calcutta - 700 032
- Maharaja Sayajirao University of Baroda, Vadodara, Gujarat
- Marathwada University, Aurangabad, Maharashtra
- Mohanlal Sukhadia University, Udaipur, Rajasthan
- National Environmental Engineering and Research Institute, Nehru Marg, Nagpur - 440 020
- National Institute of Environment and Management, Alkapuri, Kursi Road, Lucknow - 226 021
- Purvanchal University, Zafarabad Road, Jaunpur, Uttar Pradesh
- School of Environmental Studies, CUSAT, Kochi - 682 022, Kerala
- School of Environmental Studies, JNU, New Delhi - 67
- School of Planning and Architecture, IP Estate, New Delhi - 110 002
- Tamil Nadu GD Naidu Agricultural University, Coimbatore, Tamil Nadu

- University of Bombay, Fort, Mumbai - 400 001
- University of Rajasthan, Jaipur, Rajasthan

International Focus

- American Academy of Environmental Engineers 130 Holiday Court, Suite 100 Annapolis, MD 21401
- National Association of Environmental Professionals, 5165 MacArthur Boulevard, NW Washington, DC 20016-3315
- Natural Environment Research Council, Polaris House, North Star Avenue, Swindon, Wilts, SN2 1EU

11

Industrial Engineering

Industrial engineering is a branch of engineering that concerns the development, improvement, implementation and evaluation of integrated systems of people, knowledge, equipment, energy, material and process. Industrial engineering draws upon the principles and methods of engineering analysis and synthesis, as well as mathematical, physical and social sciences together with the principles and methods of engineering analysis and design to specify, predict and evaluate the results to be obtained from such systems. Industrial engineers work to eliminate wastes of time, money, materials, energy and other resources.

Industrial engineering is also known as operations management, production engineering, manufacturing engineering or manufacturing systems engineering; a distinction that seems to depend on the viewpoint or motives of the user. Recruiters or educational establishments use the names to differentiate themselves from others. In healthcare industrial engineers are more commonly known as management engineers, engineering management, or even health systems engineers.

Whereas most engineering disciplines apply skills to very specific areas, industrial engineering is applied in virtually every industry. Examples of where industrial engineering might be used include shortening lines (or queues) at a theme park, streamlining an operating room, distributing products worldwide, and manufacturing cheaper and more reliable automobiles.

The name "industrial engineer" can be misleading. While the term originally applied to manufacturing, it has grown to encompass services and other industries as well. Similar fields include operations research, systems engineering, ergonomics and quality engineering.

There are a number of things industrial engineers do in their work to make processes more efficient, to make products more manufacturable and consistent in their quality, and to increase productivity.

Areas of Expertise

The expertise required by an industrial engineer will include some or all of the following elements. People with limited education qualifications, or limited experience may specialize in only a few.

- On demand
    - o Investigate problems relating to component quality or difficulties in meeting design and method constraints.
    - o Investigate problems with the performance of processes or machines.
    - o Implement design changes at the appropriate times.
- Specifically per product (short term)
    - o Analysis of the complete product design to determine the way the whole process should be split into steps, or operations, and whether to produce sub-assemblies at certain points in the whole process. This requires knowledge of the facilities available in-house or at sub-contractors.

  - Specification of the method to be used to manufacture or assemble the product(s) at each operation. This includes the machines, tooling, jigs and fixtures and safety equipment, which may have to be designed and built. Notice may need to be taken of any quality procedures and constraints, such as ISO9000. This requires knowledge of health and safety responsibilities and quality policies. This may also involve the creation of programs for any automated machinery.
  - Measurement or calculation of the time required to perform the specified method, taking account of the skills of the operator. This is used to cost the operation performed, to allow balancing of assembly or machining flow lines or the assessment of the manufacturing capacity required. This technique is known as work study. These times are also used in value analysis.
  - Specification of the storage, handling and transportation methods and equipment required for components and finished product, and at any intermediate stages throughout the whole process. This should eliminate the possibility for damage and minimize the space required.
- Specifically per process (medium term)
  - Determine the maintenance plan for that process.
  - Assess the range of products passing through the process, then investigate the opportunities for process improvement through a reconfiguration of the existing facilities or through the purchase of more efficient equipment. This may also include the out-sourcing of that process.

This requires knowledge of design techniques and of investment analysis.

- o Review the individual products passing through the process to identify improvements that can be made by redesign of the product, to reduce (or eliminate) the cost that process adds, or to standardize the components, tooling or methods used.

- Generically (long term)
  - o Analyze the flow of products through the facilities of the factory to assess the overall efficiency, and whether the most important products have priority for the most efficient process or machine. This means maximizing throughout for the most profitable products. This requires knowledge of statistical analysis and queuing theory, and of facilities positional layout.
  - o Training of new workers in the techniques required to operate the machines or assembly processes.
  - o Project planning to achieve timely introduction of new products and processes or changes to them.
  - o Generally, a good understanding of the structure and operation of the wider elements of the Company, such as sales, purchasing, planning, design and finance; including good communication skills. Modern practice also requires good skills in participation in multi-disciplinary teams.

Value Engineering

Value engineering is based on the proposition that in any complex product, 80% of the customers need 20% of the features. By focusing on product development, one can produce a superior product at a lower cost for the major part of a market. When a customer needs more features, sell them as options. This approach is valuable in complex electromechanical products such as computer printers, in which the engineering is a major product cost.

To reduce a project's engineering and design costs, it is frequently factored into subassemblies that are designed and developed once and reused in many slightly different products. For example, a typical tape-player has a precision injection-molded tape-deck produced, assembled and tested by a small factory, and sold to numerous larger companies as a subassembly. The tooling and design expense for the tape deck is shared over many products that can look quite different. All that the other products need are the necessary mounting holes and electrical interface.

Quality Assurance and Quality Control

Quality control is a set of measures taken to ensure that defective products or services are not produced, and that the design meets performance requirements. Quality assurance covers all activities from design, development, production, installation, servicing and documentation. This field introduced the rules "fit for purpose" and "do it right the first time".

It is a truism that "quality is free." Very often, it costs no more to produce a product that always works, every time it comes off the assembly line. While this requires a conscious effort during engineering, it can considerably reduce the cost of waste and rework.

Commercial quality efforts have two foci. First, to reduce the mechanical precision needed to obtain good performance. The second is to control all manufacturing operations to ensure that every part and assembly stays within a specified tolerance. Statistical process control in manufacturing usually proceeds by randomly sampling and testing a fraction of the output. Testing every output is generally avoided due to time or cost constraints, or because it may destroy the object being tested (such as lighting matches). The variances of critical tolerances are continuously tracked, and manufacturing processes are corrected before bad parts can be produced.

A valuable process to perform on a whole consumer product is called the "shake and bake." Every so often, a whole product is mounted on a shake table in an environmental oven, and operated under increasing vibration, temperatures and humidity until it fails. This finds many unanticipated weaknesses in a product. Another related technique is to operate samples of products until they fail. Generally the data is used to drive engineering and manufacturing process improvements. Often quite simple changes can dramatically improve product service, such as changing to mold-resistant paint, or adding lock-washed placement to the training for new assembly personnel.

Many organizations use statistical process control to bring the organization to Six Sigma levels of quality. In a six sigma organization, every item that creates customer value or dissatisfaction is controlled to assure that the total number of failures are beyond the sixth sigma of likelihood in a normal distribution of customers - setting a standard for failure of fewer than four parts in one million. Items controlled often include clerical tasks such as order-entry, as well as conventional manufacturing processes.

Producibility

Quite frequently, manufactured products have unnecessary precision, production operations or parts.

Simple redesign can eliminate these, lowering costs and increasing manufacturability, reliability and profits.

For example, Russian liquid-fuel rocket motors are intentionally designed to permit ugly (though leak-free) welding, to eliminate grinding and finishing operations that do not help the motor function better.

Some Japanese disc brakes have parts toleranced to three millimeters, an easy-to-meet precision. When combined with crude statistical process controls, this assures that less than one in a million parts will fail to fit.

Many vehicle manufacturers have active programs to reduce the numbers and types of fasteners in their product, to reduce inventory, tooling and assembly costs.

Another producibility technique is near net shape forming. Often a premium forming process can eliminate hundreds of low-precision machining or drilling steps. Precision transfer stamping can quickly produce hundreds of high quality parts from generic rolls of steel and aluminum. Die casting is used to produce metal parts from aluminum or sturdy tin alloys, which are often about as strong as mild steels. Plastic injection molding is a powerful technique, especially if the special properties of the part are supplemented with inserts of brass or steel.

When a product incorporates a computer, it replaces many parts with software that fits into a single light-weight, low-power memory part or micro-controller. As computers grow faster, digital signal processing software is beginning to replace many analog electronic circuits for audio and sometimes radio frequency processing.

On some printed circuit boards, itself a producibility technique, the conductors are intentionally sized to act as delay lines, resistors and inductors to reduce the parts count. An important recent innovation was the use of "surface mounted" components. At one stroke, this eliminated the need to drill most holes in a printed circuit board, as well as clip off the leads after soldering.

In Japan, it is a standard process to design printed circuit boards of inexpensive phenolic resin and paper, and reduce the number of copper layers to one or two to lower costs without harming specifications.

It is becoming increasingly common to consider producibility in the initial stages of product design, a process referred to as design for manufacturability. It is much cheaper to consider these changes during the initial stages of design rather than redesign products after their initial design is complete.

Motion Economy

Industrial engineers study how workers perform their jobs, such as how workers or operators pick up electronic components to be placed in a circuit board or in which order the components are placed on the board. The goal is to reduce the time it takes to perform a certain job and redistribute work so as to require fewer workers for a given task.

Frederick Winslow Taylor and Frank and Lillian Gilbreth did much of the pioneering work in motion economy. Taylor's work sought to study and understand what caused workers in a coal mine to become fatigued, as well as ways to obtain greater productivity from the workers without additional man hours. The Gilbreths devised a system to categorize all movements into subgroups known as therbligs (Gilbreths spelled backwards, almost). Examples of therbligs include hold, position, and search.

Industrial engineers frequently conduct time studies

or work sampling to understand the typical role of a worker. Systems such as Maynard Operation Sequence Technique (MOST) have also been developed to understand the work content of a job.

Major Industrial Engineering Colleges in India

- Orissa Engineering College, Utkal University.
- Guru Nanak Engineering College, Ludhiana.
- BMS College of Engineering, Bangalore.
- Shri Dharmasthala Manjunatheswara College of Engineering, Dharwai.
- PES College of Engineering, Mandya.
- Regional Engineering College (REC) Warangal
- Regional Engineering College (REC) Silchar
- Bhagalpur Engineering College, Bhagalpur.
- Regional Engineering College (REC) Jamshedpur
- Regional Engineering College (REC) Surat
- Regional Engineering College (REC) Kurukshetra
- Regional Engineering College (REC) Hamirpur
- Regional Engineering College (REC) Srinagar
- Venkatesha Education Society, MVJ College of Engineering, Kadugodi, Bangalore.
- Manipal Institute of Technology.
- MS Ramaiah Institute of Technology, Bangalore.
- Dayanand Sagar College of Engineering, Bangalore.
- RV College of Engineering, Bangalore.
- Bapuji Institute of Engineering Technology, Devangere.
- KLE Engineering College, Belgaum.

- Khaja Banda Nawaz College of Engineering, Gulbarga.
- RTE Society Engineering College, Hulkoti.
- Regional Engineering College (REC) Suratkal.
- Faculty of Technology, Cochin University.
- College of Engineering, Thiruvananthapuram.
- Thangal Kunju Musaitar College of Engineering.
- Marath Nasius College of Engineering, Kothamangalam.
- Regional Engineering College (REC) Calicut
- Regional Engineering College (REC) Bhopal
- RD Kamala Nehru Engineering College, Nagpur.
- Amrutvahini College of Engineering, Sangamner, Ahmednagar
- Shyam Lal College of Engineering, Marathwada.
- Maharashtra College of Engineering Polytechnic, Marathwada.
- Bharti Vidyapeeth College of Engineering, Katraj, Pune
- College of Engineering, Pusad.
- Sir Sant Gajanan Maharaj College of Engineering, Shegaon.
- PRE Socs College of Engineering, Ahmednagar.
- Vishwakarma Institute of Technology, Pune.
- Sanjivani Education Society's College of Engineering, Kopargaon, Ahmednagar.
- Jawaharlal Nehru Engineering College, Aurangabad.
- KK Wagh Education Society, Panchavati, Nashik.

- College of Engineering, Osmanabad.
- Kolhapur Institute of Technology, Shirgaon, Kolhapur.
- College of Engineering and Technology, Laxmi Nagar, Nagpur.
- Walchand Institute of Technology, Shivaji University, Sholapur.
- Regional Engineering College (REC) Visvesvarayya, Nagpur.
- Indira Gandhi Institute of Technology, Sarang.
- Regional Engineering College (REC) Rourkela.
- Thapar Institute of Engineering and Technology, Patiala.
- Regional Engineering College (REC) Jalandhar
- Regional Engineering College (REC) Jaipur
- College of Engineering, Guindi, Chennai.
- Regional Engineering College (REC) Tiruchchirappalli
- Institute of Engineering and Technology, Aliganj, Lucknow.
- Regional Engineering College (REC) Allahabad
- Regional Engineering College (REC) Durgapur

12

Marine Engineering

Though a branch of transport engineering, Marine engineering has gained momentum these day. This provides ample career opportunities to adventurous aspirants specifically in the countries having large coastal zone.

Marine Engineers are the officers of a ship that operate and maintain the propulsion and other systems onboard the vessel. Marine Engineering staff also deal with the "Hotel" facilities onboard, notably the sewage, lighting, air conditioning and water systems. They deal with bulk fuel transfers, and require training in firefighting and first aid, as well as in dealing with the ship's boats and other nautical tasks- especially with cargo loading/ discharging gear and safety systems, though the specific cargo discharge function remains the responsibility of deck officers and deck workers.

Sometimes, they are involved in the design and construction of these complicated systems. New design is mostly included within the naval architecture or ship design. The field is closely related to mechanical engineering, although the modern engineer requires knowledge (and hands on experience) with electrical, electronic, pneumatic, hydraulic, chemistry, control engineering, naval architecture, process engineering, gas turbines and even nuclear technology on certain military vessels.

A ship's crew is divided into two distinct sections: those who navigate the ship and those who maintain it. The navigators are deck officers and report to the Master (also functionally referred to as Captain), and engineers whose charge is the running and maintenance of all machinery report to the Chief engineer. Also on board are 'ratings' who are experienced hands who, though not officers, play a key role in assisting the officers in daily operations. In the Engine department, this includes electricians, motormen (or fitters, depending on nomanclature), oilers, wipers, and utilitymen who play a large role in the daily upkeep of the engine room.

The original term engineer on a ship meant the people who dealt with the engines ("The black hand gang"), as opposed to the Consulting Engineer concept. Marine Engineers are generally much more hands on, and often get dirty, sweaty and work in arduous conditions with temperatures in the engine room spaces exceeding 50 degrees Celsius (122 degrees Fahrenheit). Care and a clear head need to be maintained at all times, however, especially with heavy machinery in a seaway, and in managing the rest of the engine-room crew.

There are 4 classifications of Marine Engineering licences throughout the merchant marine, those being Chief Engineer, First Assistant/Second Engineer, Second Assistant/Third Engineer, and Third Assistant/Fourth Engineer.

Personnel

A common Engineering crew for a ship is:

- (1) Chief Engineer
- (1) Second Engineer / First Assistant Engineer
- (1) Third Engineer / Second Assistant Engineer

- (1-2) Fourth Engineer / Third Assistant Engineer
- (0-2) Fifth Engineer / Junior Engineer
- (1-3) Oiler (unlicenced qualified rating)
- (0-3) Greaser/s (unlicensed qualified ratiyng)
- (1-5) Entry-level rating (such as Wiper, Utilityman, etc).

Insignia

Engineer Officers bear the same number of stripes on epaulettes as Deck Officers with complementary rank. For instance, a Chief Engineer bears 4 stripes, as a Captain does, but they are distinguished by the epaulettes bearing the Engineer's Blue. There is a popular legend that Engineers wear the color blue in honor of the RMS Titanic's Engineers who never abandoned their stations and went down with the ship. But that colorful story is not quite the source of the honour. In the 1830s, Joseph Whitworth invented the method of making truly accurate flat surfaces by using Engineer's blue. The method is also used for testing bearing and journal surfaces. It is this color that Engineers wear.

A different means of differentiating engineering officers from deck officers is by the device above the stripes. In some merchant fleets, a propeller device is located above the engineer's stripes while an anchor or a diamond is located above the deck officer's stripes.

Major Marine Engineering Colleges

Algeria

- Governmental—IInstitut des Sciences de la Mer et de l'Aménagement du Littoral (ISMAL)
- Governmental—Institut Supérieur Maritime (ISM) - WMU Branch

Argentina

- Governmental—Escuela Nacional de Naútica "Manuel Belgrano" - WMU Branch
- Governmental—Escuela Nacional de Pesca (National School of Fisheries)
- Governmental—Escuela Nacional Fluvial

Australia

- Governmental—Australian Maritime College
- Governmental—Fremantle Maritime Centre (SMC)
- Governmental—WA Maritime Training Centre, Challenger TAFE Fremantle Australia
- Governmental—Hunter Institute of Technology
- Governmental—Marine Technology Centre of Sydney

Bangladesh

- Governmental—Marine Academy Chittagong - WMU Branch
- Governmental—Seamen's Training Centre
- Private—International Institute of Maritime Technology (IIMT)

Belgium

- Governmental—Antwerp Maritime Academy (Nautical College)
- Private—SGS Training Institute V.Z.W.

Regional

- Governmental—Istituto Tecnico Nautico

Azerbaijan

- Governmental—Azerbaijan State Marine Academy

Brazil

- Governmental—Centro de Instruçäo "Almirante Graça Aranha" (CIAGA)
- Governmental—Centro de Instrução "Almirante Brás de Aguiar" (CIABA)

Brunei Darussalam

- Private—Megamas Training Comp. Sdn. Bhd., Integrated Safety Training Centre

Bulgaria

- Governmental—Nicola Vaptsarov Naval Academy
- Governmental—Bulgarian Maritime Training Centre (BMTC)
- Governmental—Marine College
- Governmental—Secondary Specialized School on Sea and Ocean Fishing (SSUMOR)
- Governmental—Technical University - Varna

Canada

- Governmental—Canadian Coast Guard College
- Governmental—École des Pêches du N.-B.
- Governmental—Fisheries and Marine Institute of Memorial University
- Governmental—Great Lakes International Marine Training Centre - Georgian College
- Governmental—Institut Maritime du Québec - Centre de Québec
- Governmental—Institut Maritime du Québec - Head Office
- Governmental—Marine Affairs Program, Dalhousie University
- Governmental—Marine Centre, Holland College

- Governmental—Nautical Institute - Nova Scotia Community College
- Governmental—New Brunswick Community College
- Governmental—Niagara College of Applied Arts and Technology
- Governmental—North Island College (NIC)
- Governmental—Pacific Marine Training Campus of the British Columbia Institute of Technology

Chile

- Governmental—Centro de Instrucción y Capacitación Marítima - WMU Branch
- Governmental—Escuela Naval "Arturo Prat"
- Governmental—Servicio Hidrográfico y Oceanográfico de la Armada

China

- Governmental—Dalian Marine College
- Governmental—Dalian Maritime University (DMU) - WMU Branch
- Governmental—Guangzhou Maritime College
- Governmental—Jimei Navigation Institute
- Private—Ming Wah (Shekou) Training Centre
- Governmental—Nantong Shipping College of Jiangsu Province
- Governmental—Qingdao Ocean Shipping Mariner's College
- Governmental—Shanghai Marine School
- Governmental—Shanghai Maritime University
- Governmental—Wuhan Transportation University -

Colombia

- Governmental—Centro Naútico Pesquero del Caribe - SENA
- Governmental—Escuela Naval de Cadetes "Almirante Padilla"

Côte d'Ivoire

- Regional—Académie Régionale des Sciences et Techniques de la Mer (ARSTM) - WMU Branch

Croatia

- Governmental—University of Rijeka, Faculty of Maritime Studies

Cuba

- Governmental—Academia Naval "Granma"

Cyprus

- Private—Hanseatic Marine Training School [1]
- Governmental—Higher Technical Institute (HTI)

Denmark

- Governmental—Aalborg Civilforsvar
- Governmental—Aalborg Maskinmesterskole
- Governmental—Århus Brandskole
- Governmental—Århus Maskinmesterskole
- Private—Danish Maritime Institute (DMI)
- Governmental—Danish Offshore Safety & Technology School
- Governmental—Esbjerg Brandskole
- Governmental—Esbjerg Maskinmesterskole
- Governmental—Fanø Sofartsskole
- Governmental—Fanoe Navigationsskole

- Governmental—Føroya Brandskúli (The Faroese Fire-fighting Training Centre)
- Governmental—Føroya Sjomansskuli
- Governmental—Fredericia Maskinmesterskole
- Governmental—Frederikshavn Maskinmesterskole
- Governmental—Frederikshavn Sofartsskole
- Governmental—Københavns Maskinmesterskole
- Private—Kogtved Søfartsskole
- Governmental—Korsoer Brand-og Redningsskole
- Governmental—Marstal Navigationsskole
- Private—Maersk Training Centre
- Governmental—Maskinmeistaraskulin
- Governmental—Nordjysk Branskole
- Governmental—Odense Maskinmesterskole
- Governmental—Sovaernets Havariskole
- Governmental—Søværnets Havariskole
- Governmental—Svendborg Maskinmesterskole
- Governmental—Svendborg Navigationsskole
- Private—Svendborg Sofartsskole
- Governmental—Training Ship "Danmark"
- Private—Training Ship "Georg Stage"

Egypt

- Regional—Arab Academy for Science & Technology & Maritime Transport (AASTMT) - Head Office - WMU Branch
- Regional—College of Engineering & Technology - (AASTMT)
- Regional—College of Maritime Transport & Technology - (AASTMT)

- Regional—Integrated Simulators Complex (ISC) - (AASTMT)
- Governmental—Port Training Institution (PTI)
- Regional—Cairo Maritime University

Estonia

- Governmental—Estonian Maritime Academy

Fiji

- Governmental—School of Maritime Studies

Finland

- Governmental—Åland Institute of Technology
- Governmental—Alands Sjofartslaroverk / Alands Sjomansskola
- Governmental—Helsinki Institute of Technology
- Governmental—Kotka Maritime & Transport Institute
- Private—Kotka Polytechnic
- Private—Maritime Institute of Turku
- Governmental—Rauma Maritime Institute
- Governmental—The Ship Simulation Centre
- Governmental—Turku Institute of Technology

France

- Governmental—Ecole Nationale de la Marine Marchande - Marseille
- Governmental—Ecole Nationale de la Marine Marchande - Nantes Nautical Institute
- Governmental—Ecole Nationale Supérieure de Techniques Avancées (ENSTA)
- Private—Institut Portuaire du Havre (IPER)

- Governmental—Lycée Professionnel Maritime - Le Portel
- Governmental—Lycée Professionnel Maritime - Saint Malo
- Governmental—Marins Pompiers de Marseille
- Private—Sogreah Ingenierie SNC

Georgia

- Governmental—Batumi State Maritime Academy

Germany

- Governmental—Bremerhaven Polytechnic - Technical Department FB1 (Hochschule Bremerhaven)
- Governmental—Department of Nautical Studies - Bremen University of Applied Sciences (Hochschule Bremen)
- Governmental—Fachbereich Seefahrt der FH Ostfriesland und Fachschule Seefahrt Leer
- Governmental—Fachhochschule - Flensburg
- Governmental—Fachhochschule - Oldenburg - Department of Marine Studies
- Private—Institut für Maritime Automatisierungstechnic und Navigation e. V. (MATNAV)
- Governmental—Schleswig - Holsteinische Seemannsschule (SHS)
- Governmental—Seefahrtschule Gruenendeich
- Governmental—Staatliche Seefahrtschule Cuxhaven
- Governmental—Universitat Rostock - Fakultat fur Ingenieur Wissenschaften AB Seefahrt
- Governmental—Zentrum fur Schiffsbetriebs Forschung und Ausbildung

Ghana

- Regional—Regional Maritime Academy - WMU Branch

Greece

- Governmental—Makedonia National Mercantile Marine Academy
- Governmental—Merchant Marine Centre of Further Education
- Governmental—Naval Maritime Academy for Deck Officers
- Governmental—Sukhmani Maritime Academy for Deck Eng.

Hong Kong, China

- Governmental—Haking Wong Technical Institute
- Governmental—Hong Kong Polytechnic University - Department of Maritime Studies
- Governmental—Hong Kong Technical College (Tsing Yi) - Department of Mechanical Engineering
- Governmental—The Hong Kong Sea School
- Governmental—Vocational Training Council - Seamen's Training Centre

Iceland

- Governmental—The College of Navigation (Stýrimannaskólinn í Reykjavík)
- Governmental—The Marine Engineering College of Iceland

India

- Private—Academy of Maritime Education and Training (AMET)

- Private—Anglo Eastern Maritime Training Centre
- Governmental—Cochin Shipyard Ltd. Marine Engineering Training Centre
- Private—Fosma Maritime Institute
- Private—Fosma Maritime Institute & Research Organization
- Private—GESCO. Cadet Academy
- Private—Indian Institute of Port Management
- Private—Indian Maritime College
- Governmental—L.B.S. CAMSAR (L.B.S. College of Advanced Maritime Studies and Research), Mumbai
- Private—Maersk Training Centre India
- Governmental—Marine Engineering and Research Institute, Kolkatta
- Private—Mumbai Port Trust - Fosma Seafarers' Training Centre
- Governmental—National Institute of Port Management
- Other—Orissa Maritime Academy
- Private—S.E.I. Educational Trust Sea Explorers' Institute
- Private—Seacom Marine College
- Governmental/Private—Shipping Corporation of India (SCI) - Maritime Training Institute (MTI) - WMU Branch
- Private—Sri Venkateswara College of Engineering
- Governmental—T.S. Chanakya
- Private—Training Ship "Rahaman"
- Private—(Chidambaram Institute of Maritime Technolgy)

- Private—Tolani Maritime Institute
- Governmental—[marine engg $ research institute (polyvalent degree)],mumbai

Indonesia

- Governmental—Balai Pendidikan Dan Latihan Pelayaran (BPLP), Merchant Marine Academy Semarang
- Governmental—Balai Pendidikan Dan Latihan Pelayaran (BPLP), Merchant Marine Academy Ujung Pandang
- Governmental—Balai Pendidikan Dan Latihan Pelayaran Dasar (BPLPD), Maritime Rating School, Barombong
- Governmental—Balai Pendidikan Dan Latihan Pelayaran Dasar (BPLPD), Maritime Rating School, Surabaya
- Governmental—Pendidikan Dan Latihan Ahli Pelayaran (PLAP), Marine Education and Training Institute, Jakarta

Ireland

- Governmental—BIM National Fishery Training College
- Governmental—NATIONAL MARITIME COLLEGE of IRELAND
- Private—Seas Shore Safety Services Limited

Israel

- Private—Galillee College
- Governmental/Private—Marine Training Institute

Italy

- International IMO——nternational Maritime Academy (IMA)

- Governmental—Istituto Nautico "Leone Acciaiuoli"
- Governmental—Istituto Professionale Le Di Stato Per L'Industria E Le Attivita Marinare
- Governmental—Istituto Tecnico Nautico "A. Elia"
- Governmental—Istituto Tecnico Nautico "Gioeni-Trabia"
- Governmental—Istituto Tecnico Nautico "Giorgio La Pira"
- Governmental—Istituto Tecnico Nautico "San Giorgio"
- Governmental—Istituto Tecnico Nautico Statale "Leon Palcaldo"
- Governmental—Istituto Tecnico Nautico Statale "Marino Torre"
- Governmental—Istituto Tecnico Nautico Statale "Sebastiano Venier"
- Governmental—Istituto Tecnico Nautico Statale "Tomaso di Savoia Duca di Genova"

Jamaica

- Governmental—Jamaica Maritime Institute (JMI)

Japan

- Governmental—Hiroshima National College of Maritime Technology
- Governmental—Institute for Sea Training, Ministry of Transport
- Governmental—Okinawa Kaiingakko
- Governmental—Toyama National College of Maritime Technology
- Governmental—Yuge National College of Maritime Technology

Kiribati

- Governmental—Fishery Training Centre
- Governmental—Marine Training Centre

Latvia

- Governmental—Latvian Maritime Academy
- Governmental—Latvian Shipping Company Latvian Training Centre
- Governmental—Liepaya Marine College (LMC)

Lithuania

- Governmental—Klaipeda University
- Governmental—Maritime College
- Governmental—Radar Training Centre - Lithuanian Shipping Company (LISCO)

Malawi

- Governmental—Malawi Marine Training College

Malaysia

- Private—Akademi Laut Malaysia (ALAM) - Maritime Academy of Malaysia - WMU Branch
- Private—Sribima Maritime Training Centre (SMTC)
- Governmental—Ungku Omar Polytechnic

Maldives

- Governmental—Centre for Maritime Studies - Ministry of Education

Malta

- International IMO——nternational Maritime Law Institute (IMLI)
- Private—RISC Mediterranean Fire and Safety Training School

Mexico

- Governmental—Escuela Naútica de Mazatlán
- Governmental—Escuela Naútica Mercante de Tampico - WMU Branch
- Governmental—Escuela Naútica Mercante de Veracruz "Fernando Siliceo y Torres"
- Governmental—Training Ship "Naúticas México"

Myanmar

- Governmental—Institute of Marine Technology
- Private—MTM Maritime Centre (Yangon)
- Governmental—Myanmar Shipyards Training Centre
- Private—Uniteam Marine Limited Training Centre

Netherlands

- Governmental—Berechja College - Fisheries Annex
- Private—Hogeschool Rotterdam & Omstreken. Hoger Nautisch Onderwijs
- Governmental—Hogeschool Zeeland - Maritiem Instituut "De Ruyter"
- Governmental—HVA Faculty of Technology and Maritime Science - Department of Maritime Technology
- Private—Marine Safety International Rotterdam, B.V.
- Private—Maritiem Instituut "Willem Barentsz"
- Private—Maritiem Trainingscentrum B.V. - Maritime Training Centre B.V.
- Private—Maritime Simulation Centre, The Netherlands B.V., (MSCN) Maritime Research Institute (MARIN)

- Governmental/Private—Nova College Lokatie Katwijk
- Governmental/Private—Nova College Maritiem Instituut
- Governmental—ROC Zeeland, Maritiem Instituut "De Ruyter"
- Governmental—Scheepvaart en Transport College "Meester Snyder"
- Governmental/Private—Shipping and Training College Rotterdam
- Governmental/Private—Shipping and Transport College Rotterdam / International Maritime Transport Academy (STC/IMTA)
- Governmental—Zeevaartschool 'T Noorderhooft

New Zealand

- Governmental—New Zealand Maritime School - Manukau Institute of Technology
- Governmental—New Zealand School of Fisheries, Nelson Polytechnic
- Governmental—School of Maritime Studies - Bay of Plenty Polytechnic

Nigeria

- Governmental—Department of Marine Engineering - Rivers State University of Science & Technology
- Governmental—Maritime Academy of Nigeria

Norway

- Governmental—Ålesund College (Høgskolen i Ålesund)
- Governmental—Bergen Maritime VG Skole - Bergen Maritime College

- Governmental—Bodin Nautical College
- Other—Høgskolen I Vestfold

Nigeria

- Maritime Academy of Nigeria
- Rivers State University of Science and Technology, Department of Naval Architecture and Marine Engineering

Pakistan

- Governmental—Pakistan Marine Academy (PMA) - WMU Branch

Panama

- Governmental—Escuela Naútica de Panamá

Papua New Guinea

- Governmental—Papua New Guinea Maritime College

Peru

- Governmental—Escuela Nacional de Marina Mercante "Almirante Miguel Grau" (ENAMM)

Philippines

- Private—Admiral Maritime Training Institute of the Philippines Inc. (AMTIP)
- Other—Agro Industrial Foundation College of the Philippines
- Private—Aklan Polytechnic Institute
- Private—Albatross Foundation Academy Inc.
- Private—AMOSUP Seamen's Training Center
- Private—Asian Institute of Maritime Studies (AIMS)
- Other—Baliuag Maritime Academy
- Private—Bicol Merchant Marine College

- Private—Bohol Institute of Technology
- Other—Cebu State College of Science & Technology
- Private—Central Luzon Institute of Technology (CLIT)
- Private—Colegio de la Purísima Concepción (School of the Archdiocese of Capiz)
- Other—Concord Technical Institute
- Private—Consolidated Training Systems, Inc. (CTSI) (Formerly Consolidated Maritime Resources Foundation, Inc.)
- Private—Cotabato City Central College
- Other—Davao Merchant Marine Academy
- Other—Dipolog Institute of Technology
- Other—Dr. Carlos S. Lanting College
- Other—Dr. Yanga's Francisco Balagtas College
- Other—Educational System Technological Institute
- Other—Fame Maritime Foundation Inc.
- Other—Far East Maritime Foundation Inc.
- Other—FEATI University
- Other—Golden Heritage Polytechnic College
- Other—Holy Cross of Davao College
- Other—Iloilo State College of Fisheries
- International—Institute of Technology
- Other—Inter-Global Maritime College
- Private—Inter-Island Seafarers Foundation Training Center
- Private—International Development & Environmental Shipping School (IDESS Maritime Center (Subic) Inc.)

- Private—International Maritime & Technical Institute Foundation
- Other—International Maritime and Technical Institute
- Private—John B. Lacson Colleges Foundation (JBLCF)
- Governmental—Leyte Institute of Technology
- Other—Lyceum of Batangas
- Private—Lyceum-Northwestern
- Private—Magsaysay Training Center - A Division of Magsaysay Lines Inc.
- Other—Manuel S. Enverga University
- Private—Mariner's Polytechnic College Foundation
- Private—Mariner's Polytechnic Training Center
- Other—Maritime Academy of Asia and the Pacific
- Private—Maritime Technological & Allied Services Inc. (MARITAS)
- Private—Maritime Training Center of the Philippines
- Other—MATS College of Technology
- Other—Midway Maritime and Technical Institute
- Other—Mindanao Polytechnic College
- Governmental—Misamis Institute of Technology
- Other—MTC College
- Other—NAMEI Polytechnic Institute
- Governmental—National Maritime Polytechnic
- Other—Negros Maritime College Foundation
- Governmental—Northern Iloilo Polytechnic State College

- Other—Northern Philippine College for Maritime Education & Technical Studies
- Other—Northwestern University, Philippines
- Private—Norwegian Training Center - Manila
- Private—NYK-Fil Training Center NYK-Fil Ship Management Inc.
- Other—Our Lady of Fatima College
- Private—Palawan Polytechnic College
- Other—Palompon Institute of Technology
- Other—Pamantasan ng Araulo
- Other—Pangasinan College of Science & Technology
- Private—Pangasinan Merchant Marine Academy
- Private—PHILCAMSAT (Philippine Center for Advanced Maritime Simulation and Training
- Other—Philippine Academy of Maritime Education & Technical Studies
- Other—Philippine College of Science & Technology
- Private—Philippine Institute for Maritime Studies & Technology
- Government—Philippine Merchant Marine Academy, San Narciso, Zambales
- Other—Philippine Merchant Marine School
- Other—Philippine Merchant Marine School
- Private—Philippine Nautical Training Institute
- Private—Philippine Seafarers Training Center
- Other—Philsin Marine Technological College Foundation Inc.
- Other—PMI Colleges Bohol - Tagbilaran City, Bohol, Philippines

- Other—PMI Colleges
- Governmental—Polytechnic State College of Antique
- Other—Sacred Heart School
- Private—Saint Joseph Institute of Technology
- Other—Salazar Institute of Technology
- Private—Southern de Oro Philippines College
- Other—Southern Iloilo Polytechnic
- Other—Southern Luzon College
- Other—Southwestern University
- Private—St. Therese College
- Other—Surigao Education Center
- Private—Technological Institute of the Philippines
- Other—Technological Institute of the Philippines
- Other—University of Cebu
- Other—University of Iloilo
- Other—University of Northeastern Philippines
- Other—University of Perpetual Help of Rizal
- Other—University of Perpetual Help System of Biñan
- Other—University of St. Anthony
- Private—University of the Visayas - Maritime Colleges
- Other—West Bay Colleges
- Other—West Negros College
- Other—Western Institute of Technology
- Other—Western Visayas College of Science and Technology
- Other—Zamboanga Del Sur Maritime Institute of Technology

- Other—Zamboanga State College of Marine Science and Technology

Poland

- Governmental—Akademia Marynarki Wojennej - Polish Naval Academy
- Governmental—Gdynia Maritime Academy
- Governmental—Maritime University

Republic of Korea

- Governmental—Korea Fishing Training Centre
- Governmental—Korea Marine Training & Research Institute
- Governmental—Korea Maritime University
- Governmental—Mokpo National Maritime University
- Governmental—National Fisheries University of Pusan
- Governmental—Yosu National Fisheries University

Romania

- Governmental—Marine Training Centre (MTC)
- Governmental—Merchant Marine Institute
- Governmental—Naval Academy "Mircea Cel Batran"

Russian Federation

- Governmental—Admiral Makarov State Maritime Academy (AMSMA)
- Governmental—Arkhangelsk Marine College
- Other—Astrakhan River College
- Governmental—Baltic Marine Staff Training Centre
- Private—Baltic Ships Supply Co. Limited
- Governmental—Baltic State Academy

- Private—Baltic Training Centre (BTC)
- Governmental—CNIIMF Ltd.
- Governmental—College of Water Transport
- Governmental—Educational and Training Centre of the Northen Region
- Governmental—Far Eastern Institute Refreshing for Top Executives and Specialists of Fishing Industry (DVIPK)
- Governmental—Far Eastern State Maritime Academy (FESMA)
- Governmental—Far Eastern State Technical Fisheries University
- Governmental—FESCO Training and Upgrading Courses Centre
- Governmental—GMDSS Training Centre "Primorsk Shipping Corp."
- Private—Joint/Stock North-Western Fleet
- Private—JSC "Nevelsk Marine Communications"
- Private—JSC Marine Complexes and Systems
- Governmental—Kamchaika State Maritime Academy
- Governmental—Moskow State Academy of Water Transport
- Governmental—Murnmansk Marine College
- Governmental—Murnmansk State Technical University (MSTU)
- Governmental—NORFES Co.
- Governmental—Novorassiysk State Maritime Academy
- Governmental—Prisco Training Centre

- Private—Radar Training Centre
- Private—Regional Education and Training Centre
- Governmental—Rostov-on-Don Maritime College n/a G. Sedov
- Governmental—Saint Petersburg State University of Water Communications
- Governmental—Sakhalin Marine College
- Governmental—Sakhalin Marine College
- Governmental—State Academy of Water Transport
- Private—The Training Simulating Centre of GMDSS "MARINCOM"
- Private—Training Centre of Murmansk Shipping Company
- Governmental—Training Vessel "Aviator"
- Private—TSC og GMDSS "Vosokmdreproduct" Ltd. Co.
- Governmental—Valdivostok Marine College of Fisheries
- Governmental—Volga State Academy of Water Transport

Samoa

- Governmental—Marine Training Centre

Singapore

- Governmental—National Maritime Academy - Singapore Maritime Academy, Singapore
- Governmental—Ngee Ann Polytechnic Department of Shipbuilding & Offshore Engineering
- Private—Shiphandling Simulator Centre - (PSA Institute)

- Governmental—Singapore Polytechnic
- Governmental—[Nanyang Technological University School of Civil & Environmental Engineering]

Slovenia

- Governmental—Fakulteta za Promorstvo in Promet - Univerza Ljubljani (Faculty for Maritime and Transport Studies - University of Ljubljana)
- Governmental—Srednja Pomorska Sola, "Secondary Maritime School" - Portoroz

Solomon Islands

- Governmental—Solomon Islands College of Higher Education (SICHE)

South Africa

- Governmental—Cape Technikon - Department of Maritime Studies
- Governmental—Technikon Natal - Department of Maritime Studies
- Governmental—Training Centre for Seamen
- Private—Unicorn Marine Training School - Unicorn Lines
- Governmental—Wingfield Technical College

Spain

- Governmental—Centro de Seguridad Marítima Integral
- Governmental—Centro Superior de Naútica y Estudios del Mar - Universidad de la Laguna
- Other—Colegio de Oficiales de la Marina Mercante (COMME)
- Governmental—Escola Oficial Nautico - Pesqueira

- Governmental—Escuela de Formación Profesional Naútico Pesquera - Sección Oficial del Instituto Politécnico del Estrecho
- Governmental—Escuela Reconocida de Formación Profesional Naútico Pesquera
- Governmental—Escuela Superior de la Marina Civil - Universidad de la Coruña
- Governmental—Escuela Superior de la Marina Civil - University of Cantabria
- Governmental—Escuela Superior de la Marina Civil de Bilbao - University of the Basque Country
- Governmental—Instituto de Formación Profesional Heliópolis - Departamento Marítimo - Pesquero
- Governmental—Instituto Politécnico de Formación Profesional Marítimo - Pesquero de Canarias
- Governmental—Instituto Politécnico Marítimo - Pesquero del Atlántico
- Governmental—Instituto Politécnico Marítimo - Pesquero del Mediterraneo - Alicante

Sri Lanka

- Private—Colombo International Nautical & Engineering College (CINEC)
- Private—Mercantile Seamen Training Institute (MSTI)

Sweden

- Private—Chalmers University of Technology - College of Applied Engineering and Maritime Studies
- Governmental—Kalmar University - Kalmar Maritime Academy
- Other—Sjofartens Brandskyddskommitte (Swedish Maritime Fire Protection Committee)

- International—World Maritime University (WMU)

Thailand

- Governmental—Merchant Marine Training Centre

Tonga

- Governmental—Tonga Maritime Polytechnical Institute

Trinidad and Tobago

- Governmental—Caribbean Fisheries Training and Development Institute

Tunisia

- Governmental—Academie Naval
- Governmental—Ecole de la Marine Marchande

Turkey

- Governmental—Anatolia Maritime High School
- Governmental—Faculty of Naval Architecture & Ocean Engineering - Istanbul Technical University
- Governmental—Institute of Marine Sciences and Technology - (IMST) Dokuz Eylül University
- Governmental—Maritime Faculty - Istanbul Technical University (ITUMF) - Istanbul Technical University Maritime Faculty
- Governmental—School of Maritime Business and Management (SMBM) - Dokuz Eylül University

Tuvalu

- Governmental—Tuvalu Maritime School

Ukraine

- Governmental—Kherson Maritime College of Merchant Marine Fleet

- Governmental—Odessa State Maritime Academy (OSMA)
- Governmental—Sevastopol Basic Branch Educational and Training Centre
- Governmental—Sevastopol State Technical University (SevSTU)

United Kingdom

- Private—London Institute of Shipping and Transport
- Governmental—Aberdeen College
- Private—Bournemouth University Business School
- Private—Cambridge Academy of Transport
- Governmental—Glasgow College of Nautical Studies
- Private—Jewel & Esk Valley College
- Governmental—Kilkeel College of Further Education
- Governmental—Lewis Castle College
- Governmental—London Guildhall University - Centre for International Transport Management
- Private—London Institute of Shipping and Transport
- Private—National Sea Training College
- Private—North Atlantic Fisheries College
- Private—Orkney College - Nautical Department
- Private—Southampton Institute - Warsash Maritime Centre
- Private—The Centre for Advanced Maritime Studies
- Governmental—University of Glasgow
- Governmental—University of Plymouth - Institute of Marine Studies
- Private—University of Southampton Institute of Maritime Law

- Governmental—University of Ulster
- Private—Wray Castle
- Governmental—Fleetwood Nautical Campus

United Republic of Tanzania

- Governmental—Dar Es Salaam Maritime Institute
- Governmental—Mbegani Fisheries Dev. Centre

United States of America

Governmental Academies

- Federal—United States Merchant Marine Academy
- State—California Maritime Academy
- State—Great Lakes Maritime Academy
- State—Maine Maritime Academy
- State—Massachusetts Maritime Academy
- State—State University of New York Maritime College
- State—Texas Maritime Academy

Professional Training Institutions

- Private—Calhoon M.E.B.A. Engineering School
- Private—Paul Hall Center for Maritime Training and Education
- Private—Pacific Maritime Institute
- Private—Maritime Institute of Technology and Graduate Studies
- Private—Marine Safety International
- Private—Seamen's Church Institute

Uruguay

- Governmental—Escuela Naval

Vanuatu

- Governmental—Marine Training School

Venezuela

- Governmental—Escuela de Estudios Superiores de la Marina Mercante
- Private—Fundación la Salle

Viet Nam

- Governmental—Vietnam Maritime Secondary School No.1
- Governmental—Vietnam Maritime University (VIMARU) - Haiphong

13

Materials Science and Engineering

"Whatever is not ethereal is material!"

Everything we use is made of some material or a combination of materials. We use different types of materials - metals, ceramics, polymers, semiconductors, and composites - in our daily lives. Materials Science Engineering deals with all aspects of materials:

- extraction of materials from natural and recycled resources,
- selection of appropriate materials for a given application,
- manufacturing useful objects from raw materials,
- assurance of quality of manufactured products,
- study of the behavior of materials under various conditions,
- research and development of new materials and new ways to process them into useful objects,
- analysis of the performance of materials in service and investigate causes of failure
- fundamental investigation in the relationship between materials properties and their microstructure at different length scales, and
- fundamentals of the relationship between processing and microstructure at different length scale.

In short, Materials Science & Engineering is involved with all the processes that turn natural resources into useful products that impact all facets of our economy, such as aerospace, electronics, transportation, communication, construction, recreation, entertainment, environment and energy.

Scope of Materials Science and Engineering

Materials science and engineering is an interdisciplinary field. Advances in materials enable technology progress in many fields. Historically, this connection between materials and technology has been so intimate that major periods in civilization have been named after the dominant material used in that era (e.g. Bronze Age, Iron Age). In the past few decades, at the core of the progress in such diverse fields as transportation, communication, electronics, energy and environment, are significant advances in materials. We address the scientific fundamentals of materials, their processing, and their engineering design for technological applications. We apply basic principles of chemistry, physics and biology in order to understand the structure of materials at all relevant length scales and how this structure determines its properties. We design scientific processes to fabricate materials to meet the needs of modern technology. Finally we focus on the performance of the material under specific use conditions. Perhaps, the most distinguishing feature of the discipline is the appreciation and focus on the interrelationships and interdependence between processing, structure, properties and performance.

This is a very exciting time for the field. Ever increasing performance demands in aerospace, automotive, microelectronics, environmental and biotechnology are fueling the need for materials innovation. Significant advances in our ability to characterize and manipulate materials and materials systems at different length scales are providing novel and innovative materials solutions to meet these needs. A few examples will help illustrate this point. Materials science and engineering is playing a central role in nanotechnology. The central premise of nanotechnology is exploration of phenomena and properties that manifest themselves at the nanoscale. This premise is directly aligned with the core expertise and focus of the field (interrelationship between processing, structure and properties). Photonics and optoelectronics are poised to be the next revolutionary technology for computing and communications. Development of new materials and material systems to meet the needs of this technology are proceeding at a furious pace. In biotechnology, there is need for materials solutions for longer lasting implants, scaffolds for tissue engineering, biosensors, and novel diagnostic and therapeutic strategies. New energy and environmental technologies have significant needs for materials solutions in the areas of fuel cells, pollution reduction and energy efficiency. Higher strength and stiffness materials are needed for a variety of automotive and aerospace applications. Finally, significant advances in computational materials science are providing an entirely different approach to the development of new materials and understanding their performance.

Materials scientists and engineers are employed in the manufacturing and service industries. Because of the wide ranging impact of materials on all technologies and industrial sectors, the materials science and engineering field offers a wide variety of employment opportunities. Excellent opportunities exist in all parts of the country. Materials engineers usually hold one of a variety of position titles such as metallurgist, metallurgical engineer, materials engineer/scientist, polymer engineer, ceramic engineer, manufacturing engineer, quality assurance engineer, process engineer, biomaterials scientist, corrosion engineer, and foundry engineer. The list is longer depending on the title that a particular employer attaches to the position. Employment opportunities and prospects have been very healthy. Materials scientists primarily work in research and development in academia, national labs and industrial labs. The focus of their work is on development of new materials, processes and fundamental research on the interrelationship between processing, structure and properties.

Materials scientists and engineers study the synthesis and processing of materials, their inner structure at different length scale, their properties, and their performance in machines and devices serving society. In the Materials Science & Engineering Department at University of Washington, the research and academic programs unify all classes of materials: metals, ceramics, polymers, and electronic materials. The undergraduate MSE curriculum combines general core materials subjects, taken in the sophomore and junior year, followed by advanced elective subjects which treat specific topics. Because of flexibility in choice of elective subjects, students can tailor their academic programs to suit their interests in a certain class of materials, for example, materials for the computer and electronics industry. Students receive instruction in the classroom and in the laboratory, where they learn to use research-grade equipment.

At the graduate and research level, the Department has experienced rapid expansion into new areas including polymers, hybrids, biomaterials, biomimetics, nanomaterials, photonics and magnetic materials. These new areas have applications in current and emerging industries and compliment our existing strengths in ceramics, metals, electronic materials and composites.

Basically, Materials science is an interdisciplinary field involving the properties of matter and its applications to various areas of science and engineering. It includes elements of applied physics and chemistry, as well as chemical, mechanical, civil and electrical engineering. With significant media attention to nanoscience and nanotechnology in the recent years, materials science has been propelled to the forefront at many universities, sometimes controversially.

The choice material of a given era is often its defining point: the stone age, bronze age, and steel age are examples. Materials science is one of the oldest forms of engineering and applied science. Modern materials science evolved directly from metallurgy, which itself evolved from mining. A major breakthrough in the understanding of materials occurred in the late 19th century, when Willard Gibbs demonstrated that thermodynamic properties relating to atomic structure in various phases are related to the physical properties of the material. Important elements of modern materials science are a product of the space race: the understanding and engineering of the metallic alloys and other materials that went into the construction of space vehicles was one of the enablers of space exploration. Materials science has driven, and been driven by, the development of revolutionary technologies such as plastics, semiconductors, and biomaterials.

Before the 1960s (and in some cases decades after), many *materials science* departments were named *metallurgy* departments, from a 19th and early 20th century emphasis on metals. The field has since broadened to include every class of materials, including: ceramics, polymers, semiconductors, magnetic materials, and biological materials such as medical implants.

Fundamentals of Materials Science

In materials science, rather than haphazardly looking for and discovering materials and exploiting their properties, one instead aims to understand materials fundamentally so that new materials with the desired properties can be created.

The basis of all materials science involves relating the desired properties and relative performance of a material in a certain application to the structure of the atoms and phases in that material through characterization. The major determinants of the structure of a material and thus of its properties are its constituent chemical elements and the way in which it has been processed into its final form. These, taken together and related through the laws of thermodynamics, govern the material's microstructure, and thus its properties.

An old adage in materials science says: "materials are like people; it is the defects that make them interesting". The manufacture of a perfect crystal of a material is physically impossible. Instead materials scientists manipulate the defects in crystalline materials such as precipitates, grain boundaries (Hall-Petch relationship), interstitial atoms, vacancies or substitutional atoms, creating a material with the desired properties.

Not all materials have a regular crystal structure. Polymers display varying degrees of crystallinity. Glasses, some ceramics, and many natural materials are amorphous, not possessing any long-range order in their atomic arrangements. These materials are much harder to engineer than crystalline materials. Polymers are a mixed case, and their study commonly combines elements of chemical and statistical thermodynamics to give thermodynamical, rather than mechanical descriptions of physical properties.

Besides the to industrial interest, materials science has gradually developed into a field which provides tests for condensed matter or solid state theories. New physics emerges because of the diverse new material properties needed to be explained.

Radical materials advances can drive the creation of new products or even new industries, but stable industries also employ materials scientists to make incremental improvements and troubleshoot issues with currently used materials. Industrial applications of materials science include materials design, cost-benefit tradeoffs in industrial production of materials, processing techniques (casting, rolling, welding, ion implantation, crystal growth, thin-film deposition, sintering, glassblowing, etc.), and analytical techniques (characterization techniques such as electron microscopy, x-ray diffraction, calorimetry, nuclear microscopy (HEFIB), Rutherford backscattering, neutron diffraction, etc.).

The overlap between physics and materials science has led to the offshoot field of *materials physics*, which is concerned with the physical properties of materials. The approach is generally more macroscopic and applied than in condensed matter physics. See the important publications in materials physics for more details on this field of study.

Classes of Materials (by Bond Types)

Materials science encompasses various classes of materials, each of which may constitute a separate field. Materials are sometimes classified by the type of bonding present between the atoms:

1. Ionic crystals
2. Covalent crystals
3. Metals
4. Intermetallics
5. Semiconductors
6. Polymers
7. Composite materials
8. Vitreous materials

Main Sub-fields

- Nanotechnology — rigorously, the study of materials where the effects of quantum confinement, the Gibbs-Thomson effect, or any other effect only present at the nanoscale is the defining property of the material; but more commonly, it is the creation and study of materials whose defining structural properties are anywhere from less than a nanometer to one hundred nanometers in scale, such as molecularly engineered materials.
- Crystallography — the study of how atoms in a solid fill space, the defects associated with crystal structures such as grain boundaries and dislocations, and the characterization of these structures and their relation to physical properties.
- Materials Characterization — such as diffraction with x-rays, electrons, or neutrons, and various forms of spectroscopy and chemical analysis such as Raman spectroscopy, energy-dispersive spectroscopy (EDS), chromatography, thermal analysis, electron microscope analysis, etc., in order to understand and define the properties of materials. See also List of surface analysis methods
- Metallurgy — the study of metals and their alloys, including their extraction, microstructure and processing.
- Biomaterials — materials that are derived from and/or used with biological systems.
- Electronic and magnetic materials — materials such as semiconductors used to create integrated circuits, storage media, sensors, and other devices.
- Tribology — the study of the wear of materials due to friction and other factors.

- Surface science/Catalysis — interactions and structures between solid-gas solid-liquid or solid-solid interfaces.
- Ceramics and refractories — high temperature materials including structural ceramics such as RCC, polycrystalline silicon carbide and transformation toughened ceramics

Some practitioners often consider rheology a sub-field of materials science, because it can cover any material that flows. However, modern rheology typically deals with non-Newtonian fluid dynamics, so it is often considered a sub-field of continuum mechanics. See also granular material.

- Glass Science — any non-crystalline material including inorganic glasses, vitreous metals and non-oxide glasses.

Topics that Form the Basis of Materials Science

- Thermodynamics, statistical mechanics, kinetics and physical chemistry, for phase stability, transformations (physical and chemical) and diagrams.
- Crystallography and chemical bonding, for understanding how atoms in a material are arranged.
- Mechanics, to understand the mechanical properties of materials and their structural applications.
- Solid-state physics and quantum mechanics, for the understanding of the electronic, thermal, magnetic, chemical, structural and optical properties of materials.
- Diffraction and wave mechanics, for the characterization of materials.

- Chemistry and polymer science, for the understanding of plastics, colloids, ceramics, liquid crystals, solid state chemistry, and polymers.
- Biology, for the integration of materials into biological systems.
- Continuum mechanics and statistics, for the study of fluid flows and ensemble systems.
- Mechanics of materials, for the study of the relation between the mechanical behavior of materials and their microstructures.

A short List of Non-academic Materials Facilities

Government Labs

- Argonne National Laboratory
- Lawrence Berkeley National Laboratory
- Lawrence Livermore National Laboratory
- Los Alamos National Laboratory
- Max Planck Institute
- Oak Ridge National Laboratory

Corporate Facilities

- DuPont
- GE Global Research
- IBM Thomas J. Watson Research Center

Important Journals

- Nature Materials
- Acta Materialia
- JOM
- Advanced Materials

- Computational materials science
- Advanced Functional Materials
- Journal of Materials Chemistry
- Journal of Materials Online - Open Access
- Metallurgical and Materials Transactions.

In three consecutive years—1985, 1986, and 1987—three great advances in fundamental materials science were recognized by the awarding of Nobel Prizes:

1. In 1980, in the course of fundamental studies of electrons moving in semiconductor surfaces, an entirely new and unexpected effect, the quantum Hall effect (associated with almost total absences of electrical resistance), was discovered. It should be noted that this research was certainly stimulated in part by the enormous technical interest in electrons near semiconductor surfaces. The experiments were entirely dependent on recent progress in materials science and engineering, which had made possible the preparation of surfaces with extremely well-controlled properties. The full implications of the Hall effect for our understanding of the dynamics of surface electrons are still being developed.

2 . In the early 1980s a radically new type of microscope, the scanning tunneling microscope, was developed. It depended on a subtle quantum mechanical effect, the tunneling of electrons *below* the tops of barriers (an event that could not happen under the laws of Newtonian mechanics). This technique led to incredibly accurate information about the positions of individual atoms on surfaces. Displacements of the order of 1 percent of the normal interatomic distance can be detected with the scanning tunneling microscope. Instruments of this kind can now be produced very cheaply (for about $25,000) and

are responsible for exciting advances in such fields as surface physics, electrochemistry, and biology.

3 . Superconductivity was discovered in 1911 in mercury, which lost all electrical resistance below 4.3 K above absolute zero. By the early 1970s, a painstaking materials science effort lasting more than 60 years had led to metallic compounds that remained superconducting at up to 23 K, an average increase of about 0.3 K/yr. No further progress occurred until 1986, when researchers studying entirely different classes of compounds discovered superconductivity at up to 39 K. Since then, based on accumulated knowledge of materials, compounds have been developed that remain superconducting at up to 125 K. Although major materials problems remain to be overcome (e.g., poor ductility and low critical currents), we may well be on the threshold of a technological revolution started by superconduction.

REFERENCES

Askeland, Donald R.; Pradeep P. Phulé (2005). *The Science & Engineering of Materials*, 5th edition, Thomson-Engineering.

Eberhart, Mark (2003). Why Things Break : Understanding the World by the Way It Comes Apart. Harmony.

Gaskell, David R. (1995). *Introduction to the Thermodynamics of Materials*, 4th edition, Taylor and Francis Publishing.

Gordon, James Edward (1984). The New Science of Strong Materials or Why You Don't Fall Through the Floor, Princeton University Press.

14

Mechanical Engineering

Basically, Mechanical engineering is a professional engineering discipline that involves the application of principles of physics for analysis, design, manufacturing, and maintenance of mechanical systems. It requires a solid understanding of key concepts including mechanics, kinematics, thermodynamics and energy. Practitioners of mechanical engineering, known as mechanical engineers, use these principles and others in the design and analysis of automobiles, aircraft, heating & cooling systems, buildings and bridges, industrial equipment and machinery, and more.

Prior to the Industrial Revolution, most engineering was restricted to military and civil uses. Engineers in the military, though not always referred to as such, designed fortification systems and various war machines. Civil engineers were responsible primarily for structures. "During the early 19th century in England mechanical engineering developed as a separate field to provide manufacturing machines and the engines to power them. The first British professional society of civil engineers was formed in 1818; that for mechanical engineers followed in 1847." In the America, the first mechanical engineering professional society was formed in 1880, making it the third oldest type of engineering behind civil (1852) and mining & metallurgical (1871). "The first schools in the United States to offer an engineering education were the United States Military Academy in 1817, an institution now known as Norwich University in 1819, and Rensselaer Polytechnic Institute in 1825. An engineering education is based on a strong foundation in mathematics and science; this is followed by courses emphasizing the application of this knowledge to a specific field and studies in the social sciences and humanities to give the engineer a broader education."

Education and Training

A Bachelor of Science (BS) degree in mechanical engineering is offered at many universities in the United States, and similar programs are offered at universities in most industrialized nations. In the U.S., mechanical engineering programs typically take four to five years and result in a Bachelor of Science/Arts in Mechanical Engineering, or B.S.M.E./B.A.M.E degree. Some schools offer a 4 year Bachelor of Engineering (B.E.) degree. A BE degree differ from a BS or BA degree in that the students obtain a broad based education consisting of information relevant to various engineering disciplines.

Most mechanical engineering programs are accredited nationally by ABET to ensure similar course requirements and standards between universities. The ABET web site lists 276 accredited mechanical engineering programs as of June 19, 2006.

Some mechanical engineers go on to pursue a postgraduate degree such as a Master of Engineering/ Master of Science, a Master of Engineering Management, a Doctor of Philosophy in Engineering or an Engineer's degree. The Master and Engineer's degree may consist of either research, coursework or a mixture of the two. The Doctor of Philosophy consists of a significant research component and is often viewed as the entry point to academia.

After being awarded a degree, engineers may seek licensure by a state government. To become a licensed Professional Engineer, an engineer must

- pass the comprehensive *FE (Fundamentals of Engineering)* exam,
- work a given number of years as an *Engineering Intern (EI)*,
- pass the *Principles and Practice* or *PE (Practicing Engineer or Professional Engineer)* exam.

The purpose of this process is to ensure that engineers possess the necessary technical knowledge and real-world experience to engineer safely. Some mechanical engineers choose to become licensed; those that do can be distinguished as Professional Engineers by the post-nominal title PE, as in: Jane Doe, PE. A distinction similar to practicing engineer status is the Chartered Engineer ('CEng') status awarded by some European, Asian and Oceanic engineering organizations. "In most modern countries, certain engineering tasks, such as the design of bridges, electric power plants, and chemical plants, must be approved by a Professional Engineer or a Chartered Engineer."

In India, a Bacherlor's degree in Mechanical Engineering cab be opted after 10+2.

A common entrance exam is held at state/university/IIT's level.

Course Contents

American Universities offering accredited programs in mechanical engineering are required to offer several major subjects of study, as determined by ABET (AICTE as counterpart in India). This is to ensure a minimum level of competence among graduating engineers and to inspire confidence in the engineering profession as a whole. The specific courses required to graduate, however, differ from program to program. Universities will often combine multiple subjects into a single class or split a subject into multiple classes, depending on the faculty available and the University's major area(s) of research. Fundamental subjects of mechanical engineering typically include:

- statics & dynamics
- strength of materials & solid mechanics,
- instrumentation and measurement,
- thermodynamics, heat transfer, energy conversion, and refrigeration / air conditioning,
- fluid mechanics and dynamics,
- mechanism design (including kinematics and dynamics),
- manufacturing technology or processes,
- hydraulics & pneumatics
- engineering design,
- mechatronics and/or control theory,
- drafting or CAD/CAM.

Mechanical engineers are also expected to understand and be able to apply basic concepts from chemistry, chemical engineering, electrical engineering, and physics. Most mechanical engineering programs include several semesters of calculus, as well as advanced mathematical concepts which may include differential equations and partial differential equations, linear and modern algebra, and differential geometry, among others.

Besides the core mechanical engineering curriculum, many mechanical engineering programs offer more specialized programs and classes, such as mechatronics / robotics, transport and logistics, cryogenics, fuel technology, automotive engineering, biomechanics, vibration, optics and others, if a separate department does not exist for these subjects.

Most mechanical engineering programs also require varying amounts of research or community projects to gain practical problem-solving experience. Mechanical engineering students usually hold one or more internships while studying, though this is not typically mandated by the university.

Career & Workforce Statistics

The total number of engineers employed in the U.S. in 2004 was roughly estimated to be 1.4 million. Of these, 226,000 were mechanical engineers (15.6%), second only in size to civil engineers at 237,000 (16.4%). The total number of mechanical engineering jobs in 2004 was projected to grow 9 to 17%, with average starting salaries being $50,236 with a bachelors degree, $59,880 with a masters degree, and $68,299 with a doctorate degree. This places mechanical engineering at 8th of 14 among engineering bachelors degrees, 4th of 11 among masters degrees, and 6th of 7 among doctorate degrees in average annual salary. The median annual earning of mechanical engineers in the U.S. workforce is roughly $63,000. This number is highest when working for the government ($72,500), and lowest when doing general purpose machinery manufacturing in the private sector ($55,850).

Remarkably, Canadian engineers make an average of $28.10 per hour with 3% unemployed. The average for all occupations is $16.91 per hour with 5% unemployed. Eight per cent of these engineers are self-employed, and since

1994 the proportion of female engineers has remained constant at 4%.

Mechanical Engineering Process

The process of mechanical engineering is optimization: engineers strive to optimize cost, increase productivity, durability, safety, and overall usefulness of objects. This process can be as simple as the design of a chair for comfort or as complex as the optimization of a turbocharged engine for speed. It can be as small as the cutting of a nano-sized gear or as large as the assembly of a supertanker used to carry oil around the world.

Tools and Work

Modern analysis and design processes in mechanical engineering are aided by various computational tools including FEA, CFD, and CAD/CAM. These modern processes facilitate engineers to model (create a 3D object in a computer), analyze the quality of design etc, before a prototype is created. By this the invention and experimenting with new designs becomes very easy and can be done without any money.

Major Subdisciplines

The field of mechanical engineering can be thought of as a collection of many mechanical disciplines. Many of these subdisciplines which are typically taught at the undergraduate level are listed below, with a brief explanation and the most common application of each. Some of these subdisciplines are unique to mechanical engineering, while others belong to mechanical engineering and one or more other disciplines. Most work that a mechanical engineer does uses skills and techniques from several of these subdisciplines, as well as specialized subdisciplines. Specialized subdisciplines as defined here are usually the subject of graduate more than undergraduate research.

(a) Mechanics: Mechanics is, in the most general sense, the study of forces and their effect upon matter. Typically, engineering mechanics is used to analyze and predict the acceleration and deformation (both elastic and plastic) of objects under known forces (also called loads) or stresses. Subdisciplines of mechanics include

- Statics, the study of non-moving bodies under known loads
- Dynamics (or kinetics), the study of how forces affect moving bodies
- Mechanics of materials, the study of how different materials deform under various types of stress
- Fluid Mechanics, the study of how fluids react to forces. Note that fluid mechanics can be further split into fluid statics and fluid dynamics, and is itself a subdiscipline of continuum mechanics. The application of fluid mechanics in engineering is called hydraulics.
- Continuum mechanics is a method of applying mechanics that assumes that objects are continuous. It is contrasted by discrete mechanics.

Application: Mechanical engineers typically use mechanics in the design or analysis phases of engineering. If the engineering project were the design of a vehicle, statics might be employed to design the frame of the vehicle, to evaluate where the stresses will be most intense. Dynamics might be used when designing the car's engine, to evaluate the forces in the pistons and cams as the engine cycles. Mechanics of materials might be used to choose an appropriate material for the frame or engine. Fluid mechanics might be used to design a ventilation system for the vehicle (see HVAC), or to design the intake system for the engine.

(b) Kinematics: Kinematics is the study of the motion of bodies and systems, while ignoring the forces that cause the motion. The movement of a crane and the oscillations of a piston in an engine are both simple kinematic systems. The crane is a type of open kinematic chain, while the piston is part of a closed four bar linkage.

Application: Mechanical engineers typically use kinematics in the design and analysis of mechanisms. Kinematics can be used to find the possible range of motion for a given mechanism, or, working in reverse, can be used to design a mechanism that has a desired range of motion.

(c) Mechatronics & Robotics: Mechatronics is an interdisciplinary branch of mechanical engineering, electrical engineering and software engineering that is concerned with integrating electrical and mechanical engineering to create hybrid systems. In this way, machines can be automated through the use of electric motors, servo-mechanisms, and other electrical systems in conjunction with special software. A common example of a mechatronics system is a CD-ROM drive. Mechanical systems open and close the drive, spin the CD and move the laser, while an optical system reads the data on the CD and converts it to bits. Integrated software controls the process and communicates the contents of the CD to the computer.

Application: Mechatronics is currently used in the following areas of engineering:

- Automation, and in the area of robotics.
- Servo-Mechanics
- Sensing and Control Systems
- Automotive engineering, in the design of subsystems such as anti-lock braking systems
- Computer engineering, in the design of mechanisms such as hard drives, CD-ROM drives, etc.

(d) *Robotics* is the application of mechatronics to create robots, which perform tasks that are dangerous, unpleasant, or repetitive. These robots may be of any shape and size, but all are a) preprogrammed and b) interact physically with the world. To create a robot, an engineer typically employs kinematics (to determine the robot's range of motion) and mechanics (to determine the stresses within the robot).

Application: Robots are used extensively in Industrial engineering. They allow businesses to save money on labor and perform tasks that are either too dangerous or too precise for humans to perform them economically. Many companies employ assembly lines of robots, and some factories are so reboticized that they can run by themselves. Outside the factory, robots have been employed in bomb disposal, space exploration, and many other fields. Robots are also sold residentially (see Roomba).

(e) Structural analysis: Structural analysis or just failure analysis is the branch of mechanical engineering devoted to examining why and how objects fail. Structural failures occur in two general modes: static failure, and fatigue failure. Static structural failure occurs when, upon being loaded (having a force applied) the object being analyzed either breaks or is deformed plastically, depending on the criterion for failure. Fatigue failure occurs when an object fails after a number of repeated loading and unloading cycles. Fatigue failure occurs because of imperfections in the object. A microscopic crack on the surface of the object is one type of imperfection, and it will grow slightly with each cycle (propagation) until the crack is large enough to cause failure.

Failure is not simply defined as when a part breaks, however; it is defined as when a part does not operate as intended. Some systems, such as the perforated top sections of some plastic bags, are designed to break. If these systems do not break, failure analysis might be employed to determine the cause.

Application: Structural analysis is often used by mechanical engineers after a failure has occurred, or when designing to prevent failure. Engineers may use various books and handbooks such as those published by ASM [1] to aid them in determining the type of failure and possible causes.

Structural analysis may be used the office when designing hardware, in the field to analyze failed parts, or in laboratories where parts might undergo controlled failure tests.

(f) Thermodynamics and Thermo-science: Thermodynamics is a branch of both mechanical engineering and Chemical Engineering. At its simplest, thermodynamics is the study of how energy moves through a system. Typically, engineering thermodynamics is concerned with changing energy from one form to another. Engines, for instance, change enthalpy, the stored energy in molecules, into heat and then into mechanical work that eventually turns the wheels.

Application: Thermodynamics principles are used by mechanical engineers in the fields of heat transfer, thermofluids, and energy conversion. Mechanical engineers use thermo-science to design engines and power plants, heating, ventilation, and air-conditioning (HVAC) systems, heat exchangers, heat sinks, radiators, refrigeration, insulation, and others.

(g) Drafting: Drafting or technical drawing is the mean by which mechanical engineers create instructions for manufacturing parts. A technical drawing can be a computer model or hand-drawn schematic showing all the dimensions

necessary to manufacture a part, as well as assembly notes, a list of required materials, and other pertinent information. A U.S. mechanical engineer or skilled worker who creates technical drawings may be referred to as a drafter or draftsman (or, more correctly, draftsperson). Drafting has historically been a two-dimensional process, but recent Computer-Aided Drafting (CAD) programs have begun to allow the designer to create in three dimensions.

Instructions for manufacturing a part must be fed to the necessary machinery, either manually, through programmed instructions, or through the use of a Computer-Aided Manufacturing (CAM) or combined CAD/CAM program. Optionally, an engineer may also manually manufacture a part using the technical drawings, but this is becoming an increasing rarity, except in the areas of applied spray coatings, finishes, and other processes that cannot economically be done by a machine.

Application: Drafting is used in nearly every subdiscipline of mechanical engineering, and by many other branches of engineering and architecture. Three-dimensional models created using CAD software are also commonly used in Finite element analysis (FEA) and Computational fluid dynamics (CFD).

Major Specialized Subdisciplines

The following is a list of some additional subdisciplines and topics within mechanical engineering. These topics may be considered *specialized* because they are not typically part of undergraduate mechanical engineering requirements, or require training beyond an undergraduate level to be useful.

- Acoustical Engineering
- Aerospace Engineering

- Alternative Energy
- Automotive Engineering
- Biomedical Engineering
- Computer-Aided Engineering
- Heating, Ventilation, and Air Conditioning (HVAC)
- Nanotechnology
- Nuclear Engineering
- Piping
- Power Generation
- Engineering Based Programming
- Robotics (Robotics is also listed as a general subdiscipline, but because of the breadth of the subject it may require many years of advanced training to be useful to a particular field.)

FRONTIERS OF RESEARCH IN MECHANICAL ENGINEERING

Mechanical engineering is not a static field of engineering. Mechanical engineers are constantly pushing the boundaries of what is physically possible in order to produce safer, cheaper, and more efficient machines and mechanical systems. Some technologies at the cutting edge of mechanical engineering are listed below (see also exploratory engineering).

Nanotechnology

Nanotechnology is a field of applied science and technology covering a broad range of topics. The main unifying theme is the control of matter on a scale smaller than one micrometre, as well as the fabrication of devices on this same length scale. It is a highly multidisciplinary field, drawing from fields such as colloidal science, device physics, and supramolecular chemistry. Much speculation exists as to what new science and technology might result from these lines of research. Some view nanotechnology as a marketing term that describes pre-existing lines of research.

Despite the apparent simplicity of this definition, nanotechnology actually encompasses diverse lines of inquiry. Nanotechnology cuts across many disciplines, including colloidal science, chemistry, applied physics, biology. It could variously be seen as an extension of existing sciences into the nanoscale, or as a recasting of existing sciences using a newer, more modern term. Two main approaches are used in nanotechnology: one is a "bottom-up" approach where materials and devices are built from molecular components which assemble themselves chemically using principles of molecular recognition; the other being a "top-down" approach where nano-objects are constructed from larger entities without atomic-level control.

The impetus for nanotechnology has stemmed from a renewed interest in colloidal science, coupled with a new generation of analytical tools such as the atomic force microscope (AFM) and the scanning tunneling microscope (STM). Combined with refined processes such as electron beam lithography, these instruments allow the deliberate manipulation of nanostructures, and in turn led to the observation of novel phenomena. Nanotechnology is also an umbrella description of emerging technological developments associated with sub-microscopic dimensions. Despite the great promise of numerous nanotechnologies such as quantum dots and nanotubes, real applications that have moved out of the lab and into the marketplace have mainly utilized the advantages of colloidal nanoparticles in bulk form, such as suntan lotion, cosmetics, protective coatings, and stain resistant clothing.

Fundamental Concepts

Usage of the Term

Nanotechnology is an umbrella term that is used to describe a variety of techniques to fabricate materials and devices on the nanoscale. The genesis for nanotechnology has its roots in the colloidal science of the late 19th century. These early innovations have been combined with more recent developments in device manufacture. The term has served in some regards as a means to generate new lines of funding from government agencies. One nanometer (nm) is one billionth, or 10^{-9} of a meter. For comparison, typical carbon-carbon bond lengths, or the spacing between these atoms in a molecule, are in the range .12-.15 nm, and a DNA double-helix has a diameter around 2 nm. On the other hand, the smallest cellular lifeforms, the bacteria of the genus Mycoplasma, are around 200 nm in length.

Nanotechnological techniques include those used for fabrication of nanowires, those used in semiconductor fabrication such as deep ultraviolet lithography, electron beam lithography, focused ion beam machining, nanoimprint lithography, atomic layer deposition, and molecular vapor deposition, and further including molecular self-assembly techniques such as those employing di-block copolymers. However, all of these techniques preceded the nanotech era, and are extensions in the development of scientific advancements rather than techniques which were devised with the sole purpose of creating nanotechnology or which were results of nanotechnology research.

General fields involved with proper characterization of these systems include physics, chemistry, and biology, as well as mechanical and electrical engineering. However, due to the inter- and multidisciplinary nature of nanotechnology, subdisciplines such as physical chemistry, materials science, or biomedical engineering are considered significant or essential components of nanotechnology. The design, synthesis, characterization, and application of materials are dominant concerns of nanotechnologists. The manufacture of polymers based on molecular structure, or the design of computer chip layouts based on surface science are examples of nanotechnology in modern use. Colloidal suspensions also play an essential role in nanotechnology.

Technologies currently branded with the term 'nano' are little related to and fall far short of the most ambitious and transformative technological goals of the sort in molecular manufacturing proposals, but the term still connotes such ideas. Thus there may be a danger that a "nano bubble" will form from the use of the term by scientists and entrepreneurs to garner funding, regardless of (and perhaps despite a lack of) interest in the transformative possibilities of more ambitious and far-sighted work. The above prediction has come to pass, as by 2006 over $400 million has been invested in Nanotechnology, mostly by venture capital, with very meager results. From this perspective, Nanotechnology may be viewed as a collection of wishful predictions, aimed at generating unwarranted excitement among venture capitalists.

The National Science Foundation (a major source of funding for nanotechnology in the United States) funded researcher David Berube to study the field of nanotechnology. His findings are published in the monograph "Nano-Hype: The Truth Behind the Nanotechnology Buzz". This published study (with a foreword by Mihail Roco, head of the NNI) concludes that much of what is sold as "nanotechnology" is in fact a recasting of straightforward materials science, which is leading to a "nanotech industry built solely on selling nanotubes, nanowires, and the like" which will "end up with a few suppliers selling low margin products in huge volumes."

Larger to Smaller: A Materials Perspective

A unique aspect of nanotechnology is the vastly increased ratio of surface area to volume present in many nanoscale materials which opens new possibilities in surface-based science, such as catalysis. A number of physical phenomena become noticeably pronounced as the size of the system decreases. These include statistical mechanical effects, as well as quantum mechanical effects, for example the "quantum size effect" where the electronic properties of solids are altered with great reductions in particle size. This effect does not come into play by going from macro to micro dimensions. However, it becomes dominant when the nanometer size range is reached. Additionally, a number of physical properties change when compared to macroscopic systems. One example is the increase in surface area to volume of materials. This catalytic activity also opens potential risks in their interaction with biomaterials.

Nanotechnology can be thought of as extensions of traditional disciplines towards the explicit consideration of these properties. Additionally, traditional disciplines can be re-interpreted as specific applications of nanotechnology. This dynamic reciprocation of ideas and concepts contributes to the modern understanding of the field. Broadly speaking, nanotechnology is the synthesis and application of ideas from science and engineering towards the understanding and production of novel materials and devices. These products generally make copious use of physical properties associated with small scales.

Materials reduced to the nanoscale can suddenly show very different properties compared to what they exhibit on a macroscale, enabling unique applications. For instance, opaque substances become transparent (copper); inert materials become catalysts (platinum); stable materials turn combustible (aluminum); solids turn into liquids at room temperature (gold); insulators become conductors (silicon). Materials such as gold, which is chemically inert at normal scales, can serve as a potent chemical catalyst at nanoscales. Much of the fascination with nanotechnology stems from these unique quantum and surface phenomena that matter exhibits at the nanoscale.

Nanosize powder particles (a few nanometres in diameter, also called nanoparticles) are potentially important in ceramics, powder metallurgy, the achievement of uniform nanoporosity and similar applications. The strong tendency of small particles to form clumps ("agglomerates") is a serious technological problem that impedes such applications. However, a few dispersants such as ammonium citrate (aqueous) and imidazoline or oleyl alcohol (nonaqueous) are promising additives for deagglomeration. (Dispersants are discussed in "Organic Additives And Ceramic Processing," by Daniel J. Shanefield, Kluwer Academic Publ., Boston.)

Another concern is that the volume of an object decreases as the third power of its linear dimensions, but the surface area only decreases as its second power. This somewhat subtle and unavoidable principle has huge ramifications. For example the power of a drill (or any other machine) is proportional to the volume, while the friction of the drill's bearings and gears is proportional to their surface area. For a normal-sized drill, the power of the device is enough to handily overcome any friction. However, scaling its length down by a factor of 1000, for example, decreases its power by 1000^3 (a factor of a billion) while reducing the friction by only 1000^2 (a factor of "only" a million) Proportionally it has 1000 times less power per unit friction than the original drill. If the original friction-to-power ratio was, say, 1%, that implies the smaller drill will have 10 times as much friction as power. The drill is useless.

This is why, while super-miniature electronic integrated circuits can be made to function, the same technology cannot be used to make functional mechanical devices in miniature: the friction overtakes the available power at such small scales. So while you may see microphotographs of delicately etched silicon gears, such devices are curiosities only, not actually usable parts. Surface tension increases in the same way, causing very small objects tend to stick together. This could possibly make any kind of "micro factory" impractical: even if robotic arms and hands could be scaled down, anything they pick up will tend to be impossible to put down.

All these scaling issues have to be kept in mind while evaluating any kind of nanotechnology.

Simple to Complex: A Molecular Perspective

Modern synthetic chemistry has reached the point where it is possible to prepare small molecules to almost any structure. These methods are used today to produce a wide variety of useful chemicals such as pharmaceuticals or commercial polymers. The ability of this is to extend the control to the next, seeking methods to assemble these single molecules into supramolecular assemblies consisting of many molecules arranged in a well defined manner.

These approaches utilize the concepts of molecular self-assembly and/or supramolecular chemistry to automatically arrange themselves into some useful conformation through a bottom-up approach. The concept of molecular recognition is especially important: molecules can be designed so that a specific conformation or arrangement is favored due to non-covalent intermolecular forces. The Watson-Crick basepairing rules are a direct result of this, as is the specificity of an enzyme being targeted to a single substrate, or the specific folding of the protein itself. Thus, two or more components can be designed to be complementary and mutually attractive so that they make a more complex and useful whole.

Such bottom-up approaches should, broadly speaking, be able to produce devices in parallel and much cheaper than top-down methods, but could potentially be overwhelmed as the size and complexity of the desired assembly increases. However, the bottom-up approach is viewed by many thoughtful scientists as being mostly wishful thinking. Most useful structures require complex and thermodynamically unlikely arrangements of atoms. The basic laws of probability and entropy make it very unlikely that atoms will "self-assemble" in useful configurations, or can be easily and economically nudged to do so. About the only example of this is crystal-growing, for which Nanotechnology cannot take any credit, it having been around for millenia.

Molecular Nanotechnology: A Long-term View

Advanced nanotechnology, sometimes called molecular manufacturing, is a term given to the concept of engineered nanosystems (nanoscale machines) operating on the molecular scale. By the countless examples found in biology it is currently known that billions of years of evolutionary feedback can produce sophisticated, stochastically optimized biological machines, and it is hoped that developments in nanotechnology will make possible their construction by some shorter means, perhaps using biomimetic principles. However, K Eric Drexler and other researchers have proposed that advanced nanotechnology, although perhaps initially implemented by biomimetic means, ultimately could be based on mechanical engineering principles (see also mechanosynthesis)

When the term "nanotechnology" was independently coined and popularized by Eric Drexler, who at the time was unaware of an earlier usage by Norio Taniguchi, it referred to a future manufacturing technology based on molecular machine systems. The premise was that molecular-scale biological analogies of traditional machine components demonstrated that molecular machines were possible, and that a manufacturing technology based on the mechanical functionality of these components (such as gears, bearings, motors, and structural members) would enable programmable, positional assembly to atomic specification (see the original reference PNAS-1981). The physics and engineering performance of exemplar designs were analyzed in the textbook Nanosystems.

Another view, put forth by Carlo Montemagno, is that future nanosystems will be hybrids of silicon technology and biological molecular machines, and his group's research is directed toward this end.

The seminal experiment proving that positional molecular assembly is possible was performed by Ho and Lee at Cornell University in 1999. They used a scanning tunneling microscope to move an individual carbon monoxide molecule (CO) to an individual iron atom (Fe) sitting on a flat silver crystal, and chemically bound the CO to the Fe by applying a voltage.

Though biology clearly demonstrates that molecular machine systems are possible, non-biological molecular machines are today only in their infancy. Leaders in research on non-biological molecular machines are Dr. Alex Zettl and his colleagues at Lawrence Berkeley Laboratories and UC Berkeley. They have constructed at least three distinct molecular devices whose motion is controlled from the desktop with changing voltage: a nanotube nanomotor, a molecular actuator, and a nanoelectromechanical relaxation oscillator.

Manufacturing in the context of productive nanosystems is not related to, and should be clearly distinguished from, the conventional technologies used to manufacture nanomaterials such as carbon nanotubes and nanoparticles.

There exists the potential to design and fabricate artificial structures analogous to natural cells and even organisms. Note that these are just blue-sky "potentials", and fall closer to the disciplines of Applied Biology and gene-splicing than to Nanotechnology.

Current Research

As nanotechnology is a very broad term, there are many disparate but sometimes overlapping subfields that could fall under its umbrella. The following avenues of research could be considered subfields of nanotechnology. Note that these categories are fairly nebulous and a single subfield may overlap many of them, especially as the field of nanotechnology continues to mature.

Nanomaterials

This includes subfields which develop or study materials having unique properties arising from their nanoscale dimensions.

- Colloid science has given rise to many materials which may be useful in nanotechnology, such as carbon nanotubes and other fullerenes, and various nanoparticles and nanorods.
- Nanoscale materials can also be used for bulk applications; most present commercial applications of nanotechnology are of this flavor.
- Headway has been made in using these materials for medical applications; see Nanomedicine.

Bottom-up Approaches

These seek to arrange smaller components into more complex assemblies.

- DNA Nanotechnology utilizes the specificity of Watson-Crick basepairing to construct well-defined structures out of DNA and other nucleic acids.
- More generally, molecular self-assembly seeks to use concepts of supramolecular chemistry, and molecular recognition in particular, to cause single-molecule components to automatically arrange themselves into some useful conformation.

Top-down Approaches

These seek to create smaller devices by using larger ones to direct their assembly.

- Many technologies descended from conventional solid-state silicon methods for fabricating microprocessors are now capable of creating features smaller than 100 nm, falling under the definition of nanotechnology. Giant magnetoresistance-based hard drives already on the market fit this description, as do atomic layer deposition (ALD) techniques.
- Solid-state techniques can also be used to create devices known as nanoelectromechanical systems or NEMS, which are related to microelectromechanical systems or MEMS.
- Atomic force microscope tips can be used as a nanoscale "write head" to deposit a chemical on a surface in a desired pattern in a process called dip pen nanolithography. This fits into the larger subfield of nanolithography.

Functional Approaches

These seek to develop components of a desired functionality without regard to how they might be assembled.

- Molecular electronics seeks to develop molecules with useful electronic properties. These could then be used as single-molecule components in a nanoelectronic device. For an example see rotaxane.
- Synthetic chemical methods can also be used to create synthetic molecular motors, such as in a so-called nanocar.

Speculative

These subfields seek to anticipate what inventions nanotechnology might yield, or attempt to propose an agenda along which inquiry might progress. These often take a big-picture view of nanotechnology, with more emphasis on its societal implications than the details of how such inventions could actually be created.

- Molecular nanotechnology is a proposed approach which involves manipulating single molecules in finely controlled, deterministic ways. This is more theoretical than the other subfields and is beyond current capabilities.
- Nanorobotics centers on self-sufficient machines of some functionality operating at the nanoscale.
- Programmable matter based on artificial atoms seeks to design materials whose properties can be easily and reversibly externally

Tools and Techniques

Nanoscience and nanotechnology only became possible in the 1910's with the development of the first tools to measure and make nanostructures. But the actual development started with the discovery of electrons and neutrons which showed scientists that matter can really exist on a much smaller scale than what we normally think of as small, and/or what they thought was possible at the time. It was at this time when curiosity for nanostructures had originated.

The atomic force microscope (AFM) and the Scanning Tunneling Microscope (STM) are two early versions of scanning probes that launched nanotechnology. There are other types of scanning probe microscopy, all based on the idea of the STM, that make it possible to see structures at the nanoscale. The tip of scanning probes can also be used to manipulate nanostructures (a process called positional assembly). However, this is a very slow process. This led to the development of various techniques of nanolithography such as dip pen nanolithography, electron beam lithography or nanoimprint lithography. Lithography is a top-down fabrication technique where a bulk material is reduced in size to nanoscale pattern.

The top-down approach anticipates nanodevices that must be built piece by piece in stages, much as manufactured items are currently made. Scanning probe microscopy is an important technique both for characterization and synthesis of nanomaterials. Atomic force microscopes and scanning tunneling microscopes can be used to look at surfaces and to move atoms around. By designing different tips for these microscopes, they can be used for carving out structures on surfaces and to help guide self-assembling structures. Atoms can be moved around on a surface with scanning probe microscopy techniques, but it is cumbersome, expensive and very time-consuming. For these reasons, it is not feasible to construct nanoscaled devices atom by atom. Assembling a billion transistor microchip at the rate of about one transistor an hour is inefficient. However, these techniques may eventually be used to make primitive nanomachines, which in turn can be used to make more sophisticated nanomachines.

In contrast, bottom-up techniques build or grow larger structures atom by atom or molecule by molecule. These techniques include chemical synthesis, self-assembly and positional assembly. Another variation of the bottom-up approach is molecular beam epitaxy or MBE. Researchers at Bell Telephone Laboratories like John R. Arthur. Alfred Y. Cho, and Art C. Gossard developed and implemented MBE as a research tool in the late 1960s and 1970s. Samples made by MBE were key to to the discovery of the fractional quantum Hall effect for which the 1998 Nobel Prize in Physics was awarded. MBE allows scientists to lay down atomically-precise layers of atoms and, in the process, build up complex structures. Important for research on semiconductors, MBE is also widely used to make samples and devices for the newly emerging field of spintronics.

Newer techniques such as Dual Polarisation Interferometry are enabling scientists to measure quantitatively the molecular interactions that take place at the nano-scale.

Societal Implications

Potential risks of nanotechnology can broadly be grouped into three areas:

- the risk to health and environment from nanoparticles and nanomaterials;
- the risk posed by molecular manufacturing (or advanced nanotechnology);
- societal risks.

In August 2005, a task force consisting of 50+

international experts from various fields was organized by the Center for Responsible Nanotechnology to study the societal implications of molecular nanotechnology.

In October 2005, the National Science Foundation announced that it would fund two national centers to research the potential societal implications of nanotechnology. Located at the University of California, Santa Barbara and Arizona State University, researchers at these two centers are exploring a wide range of issues including nanotechnology's historical context, technology assessment, innovation and globalization issues, and societal perceptions of risk.

A book by Geoffrey Hunt and Michael Mehta (2006) entitled *Nanotechnology: Risk, Ethics and Law* (London: Earthscan Book) provides a global overview of the state of nanotechology and society in Europe, the USA, Japan and Canada, and examines the ethics, the environmental and public health risks, and the governance and regulation of this most promising, and potentially most dangerous, of all technologies.

Determining a set of pathways for the development of molecular nanotechnology is now an objective of a broadly based technology roadmap project led by Battelle (the manager of several U.S. National Laboratories) and the Foresight Institute. That roadmap should be completed by early 2007.

Risks from Nanoparticles

The mere presence of nanomaterials (materials that contain nanoparticles) is not in itself a threat. It is only certain aspects that can make them risky, in particular their mobility and their increased reactivity. Only if certain properties of certain nanoparticles were harmful to living beings or the environment would we be faced with a genuine hazard. In this case it can be called Nanopollution.

In addressing the health and environmental impact of nanomaterials we need to differentiate two types of nanostructures: (1) Nanocomposites, nanostructured surfaces and nanocomponents (electronic, optical, sensors etc.), where nanoscale particles are incorporated into a substance, material or device ("fixed" nano-particles); and (2) "free" nanoparticles, where at some stage in production or use individual nanoparticles of a substance are present. These free nanoparticles could be nanoscale species of elements, or simple compounds, but also complex compounds where for instance a nanoparticle of a particular element is coated with another substance ("coated" nanoparticle or "core-shell" nanoparticle).

There seems to be consensus that, although one should be aware of materials containing fixed nanoparticles, the immediate concern is with free nanoparticles.

Because nanoparticles are very different from their everyday counterparts, their adverse effects cannot be derived from the known toxicity of the macro-sized material. This poses significant issues for addressing the health and environmental impact of free nanoparticles.

To complicate things further, in talking about nanoparticles it is important that a powder or liquid containing nanoparticles is almost never monodisperse, but will contain a range of particle sizes. This complicates the experimental analysis as larger nanoparticles might have different properties than smaller ones. Also, nanoparticles show a tendency to aggregate and such aggregates often behave differently from individual nanoparticles.

The lethal dose over six months for lab rats, of different kinds of nanoparticles are often characterized by a Skov Kjaer index, named after the scientist Kasper Skov Kjaer.

Health Issues

There are several potential entry routes for nanoparticles into the body. They can be inhaled, swallowed, absorbed through skin or be deliberately injected during medical procedures (or released from implants). Once within the body they are highly mobile and in some instances can even cross the blood-brain barrier.

How these nanoparticles behave inside the organism is one of the big issues that needs to be resolved. The behavior of nanoparticles is a function of their size, shape and surface reactivity with the surrounding tissue. They could cause overload on phagocytes, cells that ingest and destroy foreign matter, thereby triggering stress reactions that lead to inflammation and weaken the body's defense against other pathogens. Apart from what happens if non-degradable or slowly degradable nanoparticles accumulate in organs, another concern is their potential interaction with biological processes inside the body: because of their large surface, nanoparticles on exposure to tissue and fluids will immediately adsorb onto their surface some of the macromolecules they encounter. This may, for instance, affect the regulatory mechanisms of enzymes and other proteins.

Environmental Issues

Not enough data exists to know for sure if nanoparticles could have undesirable effects on the environment. Two areas are relevant here: (1) In free form nanoparticles can be released in the air or water during production (or production accidents) or as waste byproduct of production, and ultimately accumulate in the soil, water or plant life.

(2) In fixed form, where they are part of a manufactured substance or product, they will ultimately have to be recycled or disposed of as waste. It is not known yet whether certain nanoparticles will constitute a completely new class of non-biodegradable pollutant. In case they do, it is not known how such pollutants could be removed from air or water because most traditional filters are not suitable for such tasks (their pores are too big to catch nanoparticles).

Health and environmental issues combine in the workplace of companies engaged in producing or using nanomaterials and in the laboratories engaged in nanoscience and nanotechnology research. It is safe to say that current workplace exposure standards for dusts cannot be applied directly to nanoparticle dusts.

To properly assess the health hazards of engineered nanoparticles the whole life cycle of these particles needs to be evaluated, including their fabrication, storage and distribution, application and potential abuse, and disposal. The impact on humans or the environment may vary at different stages of the life cycle.

Regarding the risks from molecular manufacturing, an often cited worst-case scenario is "grey goo", a hypothetical substance into which the surface of the earth might be transformed by self-replicating nanobots running amok. This concept has been analyzed by Freitas in "Some Limits to Global Ecophagy by Biovorous Nanoreplicators, with Public Policy Recommendations" With the advent of nan-biotech, a different scenario called green goo has been forwarded. Here, the malignant substance is not nanobots but rather self-replicating organisms engineered through nanotechnology.

A Need for Regulation?

Regulatory bodies such as the Environmental Protection

Agency and the Food and Drug Administration in the U.S. or the Health & Consumer Protection Directorate of the European Commission have started dealing with the potential risks posed by nanoparticles. So far, neither engineered nanoparticles nor the products and materials that contain them are subject to any special regulation regarding production, handling or labeling. The Material Safety Data Sheet that must be issued for certain materials often does not differentiate between bulk and nanoscale size of the material in question and even when it does these MSDS are advisory only.

Studies of the health impact of airborne particles are the closest thing we have to a tool for assessing potential health risks from free nanoparticles. These studies have generally shown that the smaller the particles get, the more toxic they become. This is due in part to the fact that, given the same mass per volume, the dose in terms of particle numbers increases as particle size decreases.

Looking at all available data, it must be concluded that current risk assessment methodologies are not suited to the hazards associated with nanoparticles; in particular, existing toxicological and eco-toxicological methods are not up to the task; exposure evaluation (dose) needs to be expressed as quantity of nanoparticles and/or surface area rather than simply mass; equipment for routine detecting and measuring nanoparticles in air, water, or soil is inadequate; and very little is known about the physiological responses to nanoparticles.

Regulatory bodies in the U.S. as well as in the EU have concluded that nanoparticles form the potential for an entirely new risk and that it is necessary to carry out an extensive analysis of the risk. The challenge for regulators is whether a matrix can be developed which would identify nanoparticles and more complex nanoformulations which are likely to have special toxicological properties or whether it is more reasonable for each particle or formulation to be tested separately. Since null hypotheses are unfalsifiable in a true sense, nanoparticles and nanoformulations can never be proven safe. A truly precautionary approach to regulation would severely impede development in the field of nanotechnology if we require safety studies for each and every nanoscience application. Consequently, the rush seems to be on to establish a research needs assessment in the nanocommunity to preclude universal safety studies. While the outcome of these studies can form the basis for government and international regulations, a more reasonable approach might be development of a risk matrix that indentifes likely culprits.

Possible Military Applications

Societal risks from the use of nanotechnology have also been raised. On the instrumental level, these include the possibility of military applications of nanotechnology (for instance, as in implants and other means for soldier enhancement like those being developed at the Institute for Soldier Nanotechnologies at MIT) as well as enhanced surveillance capabilities through nano-sensors.

Potential Benefits and Risks for Developing Countries

Nanotechnologies may provide new solutions for the millions of people in developing countries who lack access to basic services, such as safe water, reliable energy, health care, and education. The United Nations has set Millennium Development Goals for meeting these needs. The 2004 UN Task Force on Science, Technology and Innovation noted that some of the advantages of nanotechnology include production using little labor, land, or maintenance, high productivity, low cost, and modest requirements for materials and energy.

Many developing countries, for example Costa Rica, Chile, Bangladesh, Thailand, and Malaysia, are investing considerable resources in research and development of nanotechnologies. Emerging economies such as Brazil, China, India and South Africa are spending millions of US dollars annually on R&D, and are rapidly increasing their scientific output as demonstrated by their increasing numbers of publications in peer-reviewed scientific publications. Potential opportunities of nanotechnologies to help address critical international development priorities include improved water purification systems, energy systems, medicine and pharmaceuticals, food production and nutrition, and information and communications technologies. Nanotechnologies are already incorporated in products that are on the market. Other nanotechnologies are still in the research phase, while others are concepts that are years or decades away from development.

Applying nanotechnologies in developing countries raises similar questions about the environmental, health, and societal risks described in the previous section. Additional challenges have been raised regarding the linkages between nanotechnology and development.

Protection of the environment, human health and worker safety in developing countries often suffers from a combination of factors that can include but are not limited to lack of robust environmental, human health, and worker safety regulations; poorly or unenforced regulation which is linked to a lack of physical (e.g., equipment) and human capacity (i.e., properly trained regulatory staff). Often, these nations require assistance, particularly financial assistance, to develop the scientific and institutional capacity to adequately assess and manage risks, including the necessary infrastructure such as laboratories and technology for detection.

Very little is known about the risks and broader impacts of nanotechnology. At a time of great uncertainty over the impacts of nanotechnology it will be challenging for governments, companies, civil society organizations, and the general public in developing countries, as in developed countries, to make decisions about the governance of nanotechnology.

Companies, and to a lesser extent governments and universities, are receiving patents on nanotechnology. The rapid increase in patenting of nanotechnology is illustrated by the fact that in the US, there were 500 nanotechnology patent applications in 1998 and 1,300 in 2000. Some patents are very broadly defined, which has raised concern among some groups that the rush to patent could slow innovation and drive up costs of products, thus reducing the potential for innovations that could benefit low income populations in developing countries.

There is a clear link between commodities and poverty. Many least developed countries are dependent on a few commodities for employment, government revenue, and export earnings. Many applications of nanotechnology are being developed that could impact global demand for specific commodities. For instance, certain nanoscale materials could enhance the strength and durability of rubber, which might eventually lead to a decrease in demand for natural rubber. Other nanotechnology applications may result in increases in demand for certain commodities. For example, demand for titanium may increase as a result of new uses for nanoscale titanium oxides, such as titanium dioxide nanotubes that can be used to produce and store hydrogen for use as fuel. Various organizations have called for international dialogue on mechanisms that will allow developing countries to anticipate and proactively adjust to these changes.

In 2003, Meridian Institute began the Global Dialogue on Nanotechnology and the Poor: Opportunities and Risks (GDNP) to raise awareness of the opportunities and risks of nanotechnology for developing countries, close the gaps within and between sectors of society to catalyze actions that address specific opportunities and risks of nanotechnology for developing countries, and identify ways that science and technology can play an appropriate role in the development process. The GDNP has released several publicly accessible papers on nanotechnology and development, including "Nanotechnology and the Poor: Opportunities and Risks - Closing the Gaps Within and Between Sectors of Society"; "Nanotechnology, Water, and Development"; and "Overview and Comparison of Conventional and Nano-Based Water Treatment Technologies".

Intellectual Property Issues

On the structural level, critics of nanotechnology point to a new world of ownership and corporate control opened up by nanotechnology. The claim is that, just as biotechnology's ability to manipulate genes went hand in hand with the patenting of life, so too nanotechnology's ability to manipulate molecules has led to the patenting of matter. The last few years has seen a gold rush to claim patents at the nanoscale. Over 800 nano-related patents were granted in 2003, and the numbers are increasing year to year. Corporations are already taking out broad-ranging patents on nanoscale discoveries and inventions. For example, two corporations, NEC and IBM, hold the basic patents on carbon nanotubes, one of the current cornerstones of nanotechnology. Carbon nanotubes have a wide range of uses, and look set to become crucial to several industries from electronics and computers, to strengthened materials to drug delivery and diagnostics.

Carbon nanotubes are poised to become a major traded commodity with the potential to replace major conventional raw materials. However, as their use expands, anyone seeking to manufacture or sell carbon nanotubes, no matter what the application, must first buy a license from NEC or IBM.

Nuclear Fusion

Most nuclear power plants today work on the principle of nuclear fission. An international effort is currently underway to explore the potential of nuclear fusion as a clean alternative energy source, and an experimental 500 MW power plant known as ITER is currently under construction as of 2007 in France.

In physics and nuclear chemistry, nuclear fusion is the process by which multiple nuclei join together to form a heavier nucleus. It is accompanied by the release or absorption of energy depending on the masses of the nuclei involved. Iron and nickel nuclei have the largest binding energies per nucleon of all nuclei and therefore are the most stable. The fusion of two nuclei lighter than iron or nickel generally releases energy while the fusion of nuclei heavier than iron or nickel absorbs energy; vice-versa for the reverse process, nuclear fission.

Overview

Nuclear fusion of light elements releases the energy that causes stars to shine and hydrogen bombs to explode. Nuclear fusion of heavy elements (absorbing energy) occurs in the extremely high-energy conditions of supernova explosions. Nuclear fusion in stars and supernovae is the primary process by which new natural elements are created. It is this reaction that is harnessed in fusion power.

It takes considerable energy to force nuclei to fuse, even those of the lightest element, hydrogen. But the fusion of lighter nuclei, which creates a heavier nucleus and a free neutron, will generally release more energy than it took to force them together — an exothermic process that can produce self-sustaining reactions.

The energy released in most nuclear reactions is much larger than that for chemical reactions, because the binding energy that holds a nucleus together is far greater than the energy that holds electrons to a nucleus. For example, the ionization energy gained by adding an electron to a hydrogen nucleus is 13.6 electron volts — less than one-millionth of the 17 MeV released in the D-T (deuterium-tritium) reaction shown to the top right.

Building upon the nuclear transmutation experiments of Ernest Rutherford done a few years earlier, fusion of light nuclei (hydrogen isotopes) was first observed by Mark Oliphant in 1932, and the steps of the main cycle of nuclear fusion in stars were subsequently worked out by Hans Bethe throughout the remainder of that decade.

Requirements

A substantial energy barrier must be overcome before fusion can occur. At large distances two naked nuclei repel one another because of the repulsive electrostatic force between their positively charged protons. If two nuclei can be brought close enough together, however, the electrostatic barrier can be overcome by the strong nuclear force which is stronger at close distances than the electrostatic repulsion.

When a nucleon such as a proton or neutron is added to a nucleus, the strong force attracts it to other nucleons, but primarily to its immediate neighbors due to the short range of the force. The nucleons in the interior of a nucleus have more neighboring nucleons than those on the surface. Since smaller nuclei have a larger surface area-to-volume ratio, the binding energy per nucleon due to the strong force generally increases with the size of the nucleus but approaches a limiting value corresponding to that of a fully surrounded nucleon.

The electrostatic force, on the other hand, is an inverse-square force, so a proton added to a nucleus will feel an electrostatic repulsion from all *the other protons in the nucleus. The electrostatic energy per nucleon due to the electrostatic force thus increases without limit as nuclei get larger.*

The net result of these opposing forces is that the binding energy per nucleon generally increases with increasing size, up to the elements iron and nickel, and then decreases for heavier nuclei. Eventually, the binding energy becomes negative and very heavy nuclei are not stable. The four most tightly bound nuclei, in decreasing order of binding energy, are ^{62}Ni, ^{58}Fe, ^{56}Fe, and ^{60}Ni. Even though the nickel isotope ,^{62}Ni, is more stable, the iron isotope ^{56}Fe is an order of magnitude more common. This is due to a greater disintegration rate for ^{62}Ni in the interior of stars driven by photon absorption.

A notable exception to this general trend is the helium-4 nucleus, whose binding energy is higher than that of lithium, the next heavier element. The Pauli exclusion principle provides an explanation for this exceptional behavior — it says that because protons and neutrons are fermions, they cannot exist in exactly the same state. Each proton or neutron energy state in a nucleus can accommodate both a spin up particle and a spin down particle. Helium-4 has an anomalously large binding energy because its nucleus consists of two protons and two neutrons; so all four of its nucleons can be in the ground state. Any additional nucleons would have to go into higher energy states.

The situation is similar if two nuclei are brought together. As they approach each other, all the protons in one nucleus repel all the protons in the other. Not until the two nuclei actually come in contact can the strong nuclear force take over. Consequently, even when the final energy state is lower, there is a large energy barrier that must first be overcome. It is called the Coulomb barrier.

The Coulomb barrier is smallest for isotopes of hydrogen — they contain only a single positive charge in the nucleus. A bi-proton is not stable, so neutrons must also be involved, ideally in such a way that a helium nucleus, with its extremely tight binding, is one of the products.

Using deuterium-tritium fuel, the resulting energy barrier is about 0.1 MeV. In comparison, the energy needed to remove an electron from hydrogen is 13.6 eV, about 7,500 times less energy. The (intermediate) result of the fusion is an unstable ^{5}He nucleus, which immediately ejects a neutron with 14.1 MeV. The recoil energy of the remaining ^{4}He nucleus is 3.5 MeV, so the total energy liberated is 17.6 MeV. This is many times more than what was needed to overcome the energy barrier.

If the energy to initiate the reaction comes from accelerating one of the nuclei, the process is called beam-target *fusion; if both nuclei are accelerated, it is* beam-beam *fusion. If the nuclei are part of a plasma near thermal equilibrium, one speaks of* thermonuclear *fusion. Temperature is a measure of the average kinetic energy of particles, so by heating the nuclei they will gain energy and eventually have enough to overcome this 0.1 MeV. Converting the units between electronvolts and kelvins shows that the barrier would be overcome at a temperature in excess of 1 GK, obviously a very high temperature.*

There are two effects that lower the actual temperature

needed. One is the fact that temperature is the average *kinetic energy, implying that some nuclei at this temperature would actually have much higher energy than 0.1 MeV, while others would be much lower. It is the nuclei in the high-energy tail of the velocity distribution that account for most of the fusion reactions. The other effect is quantum tunneling. The nuclei do not actually have to have enough energy to overcome the Coulomb barrier completely. If they have nearly enough energy, they can tunnel through the remaining barrier. For this reason fuel at lower temperatures will still undergo fusion events, at a lower rate.*

The reaction cross section ó is a measure of the probability of a fusion reaction as a function of the relative velocity of the two reactant nuclei. If the reactants have a distribution of velocities, e.g. a thermal distribution with thermonuclear fusion, then it is useful to perform an average of over the distributions of the product of cross section and velocity.

Fuel Confinement Methods

The fusion reaction can sustain itself if enough of the energy produced goes into keeping the fuel hot.

Gravitational

One force capable of confining the fuel well enough to satisfy the Lawson criterion is gravity. The mass needed, however, is so great that gravitational confinement is only found in stars (the smallest of which are brown dwarfs). Even if the more reactive fuel deuterium were used, a mass greater than that of the planet Jupiter would be needed.

Fuel Magnetic

Since plasmas are very good electrical conductors, magnetic fields can also confine fusion fuel. A variety of magnetic configurations can be used, the most basic distinction being between mirror confinement and toroidal confinement, especially tokamaks and stellarators.

Inertial

A third confinement principle is to apply a rapid pulse of energy to a large part of the surface of a pellet of fusion fuel, causing it to simultaneously "implode" and heat to very high pressure and temperature. If the fuel is dense enough and hot enough, the fusion reaction rate will be high enough to burn a significant fraction of the fuel before it has dissipated. To achieve these extreme conditions, the initially cold fuel must be explosively compressed. Inertial confinement is used in the hydrogen bomb, where the driver is x-rays created by a fission bomb. Inertial confinement is also attempted in "controlled" nuclear fusion, where the driver is a laser, ion, or electron beam, or a Z-pinch.

Some other confinement principles have been investigated, such as muon-catalyzed fusion, the Farnsworth-Hirsch fusor (inertial electrostatic confinement), and bubble fusion.

Production Methods

A variety of methods are known to affect nuclear fusion. Some are "cold" in the strict sense that no part of the material is hot (except for the reaction products), some are "cold" in the limited sense that the bulk of the material is at a relatively low temperature and pressure but the reactants are not, and some are "hot" fusion methods that create macroscopic regions of very high temperature and pressure.

Locally Cold Fusion

- Muon-catalyzed fusion is a well-established and

reproducible fusion process that occurs at ordinary temperatures. It was studied in detail by Steven Jones in the early 1980s. It has not been reported to produce net energy. Net energy production from this reaction is not believed to be possible because of the energy required to create muons, their 2.2 μs half-life, and the chance that a muon will bind to the new alpha particle and thus stop catalyzing fusion.

Generally Cold, Locally Hot Fusion

- Accelerator based light-ion fusion. Using particle accelerators it is possible to achieve particle kinetic energies sufficient to induce many light ion fusion reactions. Of particular relevance into this discussion are devices referred to as sealed-tube neutron generators. These small devices are miniature particle accelerators filled with deuterium and tritium gas in an arrangement which allows ions of these nuclei to be accelerated against hydride targets, also containing deuterium and tritium, where fusion takes place. Hundreds of neutron generators are produced annually for use in the petroleum industry where they are used in measurement equipment for locating and mapping oil reserves. Despite periodic reports in the popular press by scientists claiming to have invented "table-top" fusion machines, neutron generators have been around for half a century. The sizes of these devices vary but the smallest instruments are often packaged in sizes smaller than a loaf of bread. These devices do not produce a net power output.
- In sonoluminescence, acoustic shock waves create temporary bubbles that collapse shortly after creation, producing very high temperatures and pressures.

In 2002, Rusi P. Taleyarkhan reported the possibility that bubble fusion occurs in those collapsing bubbles (aka sono fusion). As of 2005, experiments to determine whether fusion is occurring give conflicting results. If fusion is occurring, it is because the local temperature and pressure are sufficiently high to produce hot fusion.

- The Farnsworth-Hirsch Fusor is a tabletop device in which fusion occurs. This fusion comes from high effective temperatures produced by electrostatic acceleration of ions. The device can be built inexpensively, but it too is unable to produce a net power output.
- Antimatter-initialized fusion uses small amounts of antimatter to trigger a tiny fusion explosion. This has been studied primarily in the context of making nuclear pulse propulsion feasible. This is not near becoming a practical power source, due to the cost of manufacturing antimatter alone.
- Pyroelectric fusion was reported in April 2005 by a team at UCLA. The scientists used a pyroelectric crystal heated from "34 to 7°C ("30 to 45°F), combined with a tungsten needle to produce an electric field of about 25 gigavolts per meter to ionize and accelerate deuterium nuclei into an erbium deuteride target. Though the energy of the deuterium ions generated by the crystal has not been directly measured, the authors used 100 keV (a temperature of about 10^9 K) as an estimate in their modeling. At these energy levels, two deuterium nuclei can fuse together to produce a helium-3 nucleus, a 2.45 MeV neutron and bremsstrahlung. Although it makes a useful neutron generator, the apparatus is not intended for power generation since it requires far more energy than it produces.

Hot Fusion

"Standard" "hot" fusion, in which the fuel reaches tremendous temperature and pressure inside a fusion reactor or nuclear weapon.

The methods in the second group are examples of non-equilibrium systems, in which very high temperatures and pressures are produced in a relatively small region adjacent to material of much lower temperature. In his doctoral thesis for MIT, Todd Rider did a theoretical study of all quasineutral, isotropic, non-equilibrium fusion systems. He demonstrated that all such systems will leak energy at a rapid rate due to bremsstrahlung, radiation produced when electrons in the plasma hit other electrons or ions at a cooler temperature and suddenly decelerate. The problem is not as pronounced in a hot plasma because the range of temperatures, and thus the magnitude of the deceleration, is much lower. Note that Rider's work does not apply to non-neutral and/or anisotropic non-equilibrium plasmas.

Important Reactions

Astrophysical Reaction Chains

The most important fusion process in nature is that which powers the stars. The net result is the fusion of four protons into one alpha particle, with the release of two positrons, two neutrinos (which changes two of the protons into neutrons), and energy, but several individual reactions are involved, depending on the mass of the star. For stars the size of the sun or smaller, the proton-proton chain dominates. In heavier stars, the CNO cycle is more important. Both types of processes are responsible for the creation of new elements as part of stellar nucleosynthesis.

At the temperatures and densities in stellar cores the rates of fusion reactions are still extremely slow. For example, at solar core temperature (T ~ 15 MK) and density

(~120 g/cm^3), the energy release rate is only ~0.1 microwatt/cm^3 — millions of times less than the rate of energy release of ordinary candela and thousands of times less than the rate at which a human body generates heat. Thus, reproduction of stellar core conditions in a lab for nuclear fusion power production is completely impractical.

Criteria and Candidates for Terrestrial Reactions

In man-made fusion, the primary fuel is not constrained to be protons and higher temperatures can be used, so reactions with larger cross-sections are chosen. This implies a lower Lawson criterion, and therefore less startup effort. Another concern is the production of neutrons, which activate the reactor structure radiologically, but also have the advantages of allowing volumetric extraction of the fusion energy and tritium breeding. Reactions that release no neutrons are referred to as *aneutronic*. In order to be useful as a source of energy, a fusion reaction must satisfy several criteria. It must:

- be exothermic - This may be obvious, but it limits the reactants to the low Z (number of protons) side of the curve of binding energy. It also makes helium He-4 the most common product because of its extraordinarily tight binding, although He3 and H3 also show up
- involve low Z nuclei - This is because the electrostatic repulsion must be overcome before the nuclei are close enough to fuse
- have two reactants - At anything less than stellar densities, three body collisions are too improbable. It should be noted that in inertial confinement, both stellar densities and temperatures are exceeded to compensate for the shortcomings of the third parameter of the Lawson criterion, ICF's very short confinement time.

- have two or more products - This allows simultaneous conservation of energy and momentum without relying on the electromagnetic force
- conserve both protons and neutrons - The cross sections for the weak interaction are too small

Neutronicity, Confinement Requirement, and Power Density

Any of the reactions above can in principle be the basis of fusion power production. In addition to the temperature and cross section discussed above, we must consider the total energy of the fusion products E_{fus}, the energy of the charged fusion products E_{ch}, and the atomic number Z of the non-hydrogenic reactant.

Specification of the D-D reaction entails some difficulties, though. To begin with, one must average over the two branches (2) and (3). More difficult is to decide how to treat the T and ^{3}He products. T burns so well in a deuterium plasma that it is almost impossible to extract from the plasma. The D-^{3}He reaction is optimized at a much higher temperature, so the burnup at the optimum D-D temperature may be low, so it seems reasonable to assume the T but not the ^{3}He gets burned up and adds its energy to the net reaction. Thus we will count the DD fusion energy as $E_{fus} = (4.03+17.6+3.27)/2 = 12.5$ MeV and the energy in charged particles as $E_{ch} = (4.03+3.5+0.82)/2 = 4.2$ MeV.

Another unique aspect of the D-D reaction is that there is only one reactant, which must be taken into account when calculating the reaction rate.

The last column is the neutronicity of the reaction, the fraction of the fusion energy released as neutrons. This is an important indicator of the magnitude of the problems associated with neutrons like radiation damage, biological shielding, remote handling, and safety. For the first two reactions it is calculated as $(E_{fus}$-$E_{ch})/E_{fus}$. For the last two reactions, where this calculation would give zero, the values quoted are rough estimates based on side reactions that produce neutrons in a plasma in thermal equilibrium.

Of course, the reactants should also be mixed in the optimal proportions. This is the case when each reactant ion plus its associated electrons accounts for half the pressure. Assuming that the total pressure is fixed, this means that density of the non-hydrogenic ion is smaller than that of the hydrogenic ion by a factor $2/(Z+1)$. Therefore the rate for these reactions is reduced by the same factor, on top of any differences in the values of <óv>/T^2. On the other hand, because the D-D reaction has only one reactant, the rate is twice as high as if the fuel were divided between two hydrogenic species.

Thus there is a "penalty" of (2/(Z+1)) for non-hydrogenic fuels arising from the fact that they require more electrons, which take up pressure without participating in the fusion reaction. (It is usually a good assumption that the electron temperature will be nearly equal to the ion temperature. Some authors, however discuss the possibility that the electrons could be maintained substantially colder than the ions. In such a case, known as a "hot ion mode", the "penalty" would not apply.) There is at the same time a "bonus" of a factor 2 for D-D due to the fact that each ion can react with any of the other ions, not just a fraction of them.

The maximum value of <óv>/T^2 is taken from a previous table. The "penalty/bonus" factor is that related to a non-hydrogenic reactant or a single-species reaction. The values in the column "reactivity" are found by dividing 1.24×10^{-24} by the product of the second and third columns. It indicates the factor by which the other reactions occur more slowly than the D-T reaction under comparable conditions. The column "Lawson criterion" weights these results with E_{ch} and gives an indication of how much more difficult it is to achieve ignition with these reactions, relative to the difficulty for the D-T reaction. The last column is labeled "power density" and weights the practical reactivity with E_{fus}. It indicates how much lower the fusion power density of the other reactions is compared to the D-T reaction and can be considered a measure of the economic potential.

Bremsstrahlung Losses in Quasineutral, Osotropic Plasmas

The ions undergoing fusion in many systems will essentially never occur alone but will be mixed with electrons that neutralize the ions' electrical charge and form a plasma. The electrons will generally have a temperature comparable to or greater than that of the ions, so they will collide with the ions and emit Bremsstrahlung. The Sun and stars are opaque to Bremsstrahlung, but essentially any terrestrial fusion reactor will be optically thin at relevant wavelengths. Bremsstrahlung is also difficult to reflect and difficult to convert directly to electricity, so the ratio of fusion power produced to Bremsstrahlung radiation lost is an important figure of merit. This ratio is generally maximized at a much higher temperature than that which maximizes the power density (see the previous subsection). The following table shows the rough optimum temperature and the power ratio at that temperature for several reactions.

fuel	T_i *(keV)*	$P_{fusion}/P_{Bremsstrahlung}$
D-T	50	140
D-D	500	2.9
D-^{3}He	100	5.3
^{3}He-^{3}He	1000	0.72
p-^{6}Li	800	0.21
p-^{11}B	300	0.57

The actual ratios of fusion to Bremsstrahlung power will likely be significantly lower for several reasons. For one, the calculation assumes that the energy of the fusion products is transmitted completely to the fuel ions, which then lose energy to the electrons by collisions, which in turn lose energy by Bremsstrahlung. However because the fusion products move much faster than the fuel ions, they will give up a significant fraction of their energy directly to the electrons. Secondly, the plasma is assumed to be composed purely of fuel ions. In practice, there will be a significant proportion of impurity ions, which will lower the ratio. In particular, the fusion products themselves *must* remain in the plasma until they have given up their energy, and *will* remain some time after that in any proposed confinement scheme. Finally, all channels of energy loss other than Bremsstrahlung have been neglected. The last two factors are related. On theoretical and experimental grounds, particle and energy confinement seem to be closely related. In a confinement scheme that does a good job of retaining energy, fusion products will build up. If the fusion products are efficiently ejected, then energy confinement will be poor, too.

The temperatures maximizing the fusion power compared to the Bremsstrahlung are in every case higher than the temperature that maximizes the power density and minimizes the required value of the fusion triple product. This will not change the optimum operating point for D-T very much because the Bremsstrahlung fraction is low, but it will push the other fuels into regimes where the power density relative to D-T is even lower and the required confinement even more difficult to achieve. For D-D and D-^{3}He, Bremsstrahlung losses will be a serious, possibly prohibitive problem. For ^{3}He-^{3}He, p-^{6}Li and p-^{11}B the Bremsstrahlung losses appear to make a fusion reactor using these fuels with a quasineutral, anisotropic plasma impossible. Some ways out of this dilemma are considered — and rejected — in *Fundamental limitations on plasma fusion systems not in thermodynamic equilibrium* by Todd Rider. This limitation does not apply to non-neutral and anisotropic plasmas; however, these have their own challenges to contend with.

REFERENCES

ABET searchable database of accredited engineering programs, Accessed *June 19, 2006*

Answers.com Engineering Article - http://www.answers.com/topic/engineering - This article references: the Engineering article in The Columbia Electronic Encyclopedia, Sixth Edition Copyright © 2003, Columbia University Press.

BBC News report on ITER - http://news.bbc.co.uk/1/hi/sci/tech/4629239.stm - Accessed *19 June 2006*

Harvard Mechanical Engineering Page - http://www.deas.harvard.edu/undergradstudy/engineeringsciences/mechanical/index.html - Accessed *19 June 2006*

http://www.jobfutures.ca/noc/2132p4.shtml - Accessed *June 19, 2006*

http://www.worldwidelearn.com/online-education-guide/engineering/mechanical-engineering-major.htm - Website cites NACE and Dept. of Labor as sources, but was unable to verify. Accessed *19 June 2006*

MIT Engineering Electives - http://www-me.mit.edu/UGradProgram/MERequirements.htm Accessed *19 June 2006*

Types of post-graduate degrees offered at MIT - http://www-me.mit.edu/GradProgram/GradDegrees.htm - Accessed *19 June 2006*

U.S. Department of Labor, Bureau of Labor Statistics, Engineering - http://www.bls.gov/oco/ocos027.htm#earnings - Accessed *19 June 2006*

University of Tulsa Required ME Courses - http://www.me.utulsa.edu/Undergraduate.html - Accessed *19 June 2006*

15

Mechatronics Engineering

Basically, Mechatronics is the synergistic combination of mechanical engineering ("mecha" for mechanisms, i.e., machines that 'move'), electronic engineering ("tronics" for electronics), and software engineering. The purpose of this interdisciplinary engineering field is the study of automata from an engineering perspective and serves the purposes of controlling advanced hybrid systems.

Mechatronics is centred on mechanics, electronics and computing which, combined, make possible the generation of simpler, more economical, reliable and versatile systems. The portmanteau "mechatronics" was first coined by Mr. Tetsuro Mori, a senior engineer of the Japanese company Yaskawa, in 1969. Mechatronics may alternatively be referred to as "electromechanical systems" or less often as "control and automation engineering".

Engineering cybernetics deals with the question of control engineering of mechatronic systems. It is used to control or regulate such a system; see control theory. Through collaboration the mechatronic modules perform the production goals and inherit flexible and agile manufacturing properties in the production scheme. Modern production equipment consists of mechatronic modules that are integrated according to a control architecture. The most known architectures involve hierarchy, polyarchy, heterarchy and hybrid. The methods for achieving a technical effect are described by control algorithms, which may or may not utilize formal methods in their design. Hybrid-systems important to Mechatronics include production systems, synergy-drives, planetary-rovers, automotive subsystems such as anti-lock braking systems, spin-assist and every day equipment such as autofocus cameras, video, hard disks, CD-players, washing machines, lego-matics etc.

A typical mechatronic engineering degree would involve classes in engineering mathematics, mechanics, machine component design, mechanical design, thermodynamics, circuits and systems, electronics and communications, control theory, digital signal processing, power engineering, robotics and usually a final year thesis.

Variant of this Field

An emerging variant of this field is biomechatronics, *whose purpose is to integrate machine and man, usually in the form of removable gadgets such as exoskeleton. This is the "real-life" version of cyberware.*

REFERENCES

Bradley, Dawson et al, *Mechatronics, Electronics in products and processes*, Chapman and Hall Verlag, London, 1991.

Karnopp, Dean C., Donald L. Margolis, Ronald C. Rosenberg, *System Dynamics: Modeling and Simulation of Mechatronic Systems*, 4th Edition, Wiley, 2006. ISBN 0-471-70965-4 Bestselling system dynamics book using bond graph approach.

Mechatronics Principles: Concepts and Applications, Tata McGraw-Hill

16

Metallurgy Engineering

Basically, Metallurgy is a domain of materials science and of materials engineering that studies the physical and chemical behavior of metallic elements, their intermetallic compounds, and their mixtures, which are called alloys.

Extractive metallurgy is the practice of separating metals from their ore, and refining them into a pure metal. In order to convert a metal oxide or sulfide to a metal, the ore must be reduced either chemically or electrolytically.

In production engineering, metallurgy is concerned with the production of metallic components for use in consumer or engineering products. This involves the production of alloys, the shaping, the heat treatment and the surface treatment of the product. The task of the metallurgist is to achieve design criteria specified by the mechanical engineer, such as cost, weight, strength, toughness, hardness, corrosion and fatigue resistance, and performance in temperature extremes.

Common engineering metals are aluminium, chromium, copper, iron, magnesium, nickel, titanium and zinc. These are most often used as alloys. Much effort has been placed on understanding one very important alloy system, that of purified iron, which has carbon dissolved in it, better known as steel. Normal steel is used in low cost, high strength applications where weight and corrosion are not a problem. Cast irons, including ductile iron are also part of this system.

Stainless steel or galvanized steel are used where resistance to corrosion is important. Aluminium alloys and magnesium alloys are used for applications where strength and lightness are required.

Most engineering metals are stronger than most plastics and are tougher than most ceramics. Composites of plastics and materials such as glass fibre and carbon fibre rival metals in applications requiring high tensile strength with little weight. Concrete rivals metals in applications requiring high compressive strength and resistance to the effects of water. Wood rivals metal in applications requiring low cost and availability of materials and low cost of construction, as well as in applications requiring certain aesthetics.

The operating environment of the product is very important; a well-designed material will resist expected failure modes such as corrosion, stress concentration, metal fatigue, creep and environmental stress fracture. Ferrous metals and some aluminium alloys in water and especially in an electrolytic solution such as seawater, corrode quickly. Metals in cold or cryogenic conditions tend to lose their toughness becoming more brittle and prone to cracking. Metals under continual cyclic loading can suffer from metal fatigue. Metals under constant stress in hot conditions can creep.

Production Engineering of Metals

Metals are shaped by processes such as casting, forging, rolling, extrusion, sintering, metalworking, machining and fabrication. With casting, molten metal is poured into a shaped mould. With forging, a red-hot billet is hammered into shape. With rolling, a billet is passed through successively narrower rollers to create a sheet. With extrusion, a hot and malleable metal is forced under pressure through a die, which shapes it before it cools. With sintering, a powdered metal is compressed into a die at high temperature. With machining, lathes, milling machines, planing machines and drills are used to cut the cold metal to shape. With fabrication, sheets of metal are cut with guillotines or gas cutters and bent into shape.

"Cold working" processes, such as rolling and fabrication, where the product's shape is altered while the product is cold, can increase the strength of the product by a process called work hardening. Work hardening creates microscopic defects in the metal, which resist further changes of shape.

Various forms of casting exist in industry and academia. These include sand casting, investment casting (also called the "lost wax process"), die casting and continuous casting.

Welding is a technique for joining certain ferrous metals and certain aluminium alloys. The metals in the weld and on both sides of the join are generally similar alloys. Brazing is a technique for joining copper-based metals.

Metals can be heat-treated by annealing, quenching, tempering and case hardening to alter properties of toughness, hardness or resistance to corrosion. Annealing is used to make a shaped product tougher by reducing the effects of work hardening, it also softens the metal. Quenching and case hardening are used to make a shaped product harder. Quenching by it self makes the metal very hard and very brittle, tempering after quenching is used to reduce the brittleness and improve overall properties.

Electroplating is the main surface treatment technique and involves bonding a thin layer of another protective metal such as gold, silver, chromium or zinc to the surface of the product to reduce corrosion.

Electrical and Electronic Engineering

Metallurgy is also applied to electrical and electronic materials where as metals such as aluminium, copper, tin and gold are used in power lines, wires, printed circuit boards and integrated circuits.

Soldering is a method of joining metallic electrical conductors where high strength is not required.

Metallurgical Techniques

Metallurgists study the microscopic and macroscopic mechanisms that cause a metal or alloy to behave in the way that it does, i.e. the changes that occur on the atomic level that affect the metal's (or alloy's) macroscopic properties. Examples of tools used for microscopic examination of metals are optical and electron microscopes and mass spectrometers.

Metallurgists study crystallography, the effects of temperature and heat treatment on the component phases of alloys, such as the eutectic and the properties of those alloy phases.

The macroscopic properties of metals are tested using machines and devices that measure tensile strength, compressive strength and hardness.

17

Mining Engineering

The need for mineral extraction and production is an essential activity of any technically proficient society. As minerals are produced from within a naturally occurring environment, disturbance of the environment as a result of mineral production is a given. Modern mining engineers must therefore be concerned not only with the production and processing of mineral commodities, but also with the mitigation of damage or changes to an environment as a result of that production and processing.

The two primary types of mine are underground mines and open-pit mines. Minerals that exist mostly underground (eg. coal, gold etc.) are generally recovered using the underground mining process. Minerals like iron ore, limestone, manganese ore, etc. are mostly recovered from the surface downwards in opencast mining.

Mining Engineering is a field that involves many of the other engineering disciplines as applied to extracting and processing minerals from a naturally occurring environment.

Engineering disciplines that are closely related to mining engineering are:

- Civil engineering;

- Environmental engineering;
- Geotechnical engineering;
- Hydraulic engineering; and
- Electrical engineering.
- Structural Engineering.

Specialized areas of mining engineering involve extraction of minerals from underwater mines, seawater, *in-situ* retorting of rock, and underground gasification.

Blasting

Explosions used to break up a rock formation and aid in the collection of ore is called blasting. There are two types of blasting: high velocity and low velocity. High velocity blasting uses explosives that have high rates of reactions and produce high pressures (i.e. high explosives). Low velocity blasting is done with explosives which have a low rate of reaction and thus low pressures (i.e. low explosives). Blasting is done in selected regions where the ore is available. The size of the ore after blasting varies.

Mineral Prospecting

Mining operations start only after geophysical survey. A geological survey will develop an estimation of the size and purity of an ore deposit. Based on these projections, more detailed studies may be ordered. This is known as detailed prospecting. One of the methods of prospecting is to extract core samples from the area of the potential mine. The cross sections of these cores can be used to interpolate a three-dimensional model of the mineral deposit.

School of Mines

There exist many institutions called School of Mines, generally engineering schools originating in 18th or 19th century industrialization. These schools no longer necessarily teach mining-related topics.

Schools of Mining

- The Western Australia School of Mines in Kalgoorlie, Australia
- The Bendigo School of Mines in Bendigo, Victoria, Australia
- The Colorado School of Mines in Colorado, United States
- The Camborne School of Mines in Cornwall, UK
- The Royal School of Mines in London, UK
- The Indian School of Mines in Dhanbad, India

Universities Teaching Mining

- Universidade Federal de MInas Gerais, Department of mining engineering in Belo Horizonte, Minas Gerais, Brazil
- The Missouri School of Mines and Metallurgy - now known as the University of Missouri–Rolla in Rolla, Missouri
- Queen's University, Kingston, Ontario, Canada
- Universidad de Santiago de Chile, Departamento de Ingeniería en Minas, Santiago, Chile
- Centro de Minería, Pontificia Universidad Católica de Chile in Santiago, Chile
- The University of Tehran, Department of Mining Engineering in Tehran, Iran
- South Dakota School of Mines and Technology in South Dakota, United States
- Michigan Technological University

- Faculty of Mining and Geology, Silesian University of Technology in Gliwice, Poland
- the *École nationale supérieure des Mines de Paris* in Paris, France
- The Welsh School of Mines, in Pontypridd, Glamorgan - now the University of Glamorgan
- Laurentian University, Sudbury, Ontario, Canada
- The *Instituto Superior Técnico da Universidade Técnica de Lisboa*, in Lisbon, Portugal
- Virginia Polytechnic Institute and State University, in Blacksburg, VA, United States
- West Virginia University, in Morgantown, WVA, United States
- University of Utah, Department of Mining Engineering in Salt Lake City, Utah, United States
- University of Alberta, Department of Mining Engineering in Edmonton, Alberta, Canada
- University of Nevada, Reno, Department of Mining Engineering in Reno, Nevada, United States
- University of New South Wales, School of Mining Engineering
- Escola de Minas de Ouro Preto(*EMOP*) Minas Gerais, Brazil. It started as an isolated school of mines. When the Universidade Federal de Ouro Preto(*UFOP*), was founded in this city, the School of Mines acted as the seed for the expansion of UFOP.
- Escuela Técnica Superior de Ingenieros de Minas de Madrid, Universidad Politécnica de Madrid, Madrid, Spain

18

Nuclear Engineering

Nuclear engineering is the practical application of the atomic nucleus gleaned from principles of nuclear physics and the interaction and maintenance of nuclear fission systems and components, specifically, nuclear reactors, nuclear power plants and/or nuclear weapons. The field can also include the study of nuclear fusion, medical applications of radiation, nuclear safety, heat transport, nuclear fuels technology, nuclear proliferation, and the effect of radioactive waste or radioactivity in the environment.

Training

The following is the typical coursework included in most U.S. nuclear engineering degree programs.

College Preparation

As with any engineering discipline, college preparation should include mathematics training through the beginnings of calculus, as well as introductory courses in physics and chemistry.

Undergraduate Coursework

Undergraduate coursework should begin with a foundation in mechanics and dynamics of particle motion, thermodynamics, introductory computer programming, college level physics and chemistry, and a rigorous training in mathematics through differential equations.

Midway through undergraduate training a nuclear engineer must choose a specialization within his field that he will further study. Further coursework in a nuclear engineering program includes but is not limited to fluid mechanics, reactor physics, quantum mechanics, thermal hydraulics, linear circuits, radiation effects, and neutron transport.

Specialization in fission, includes the study of nuclear reactors, fission systems, and nuclear power plants, the primary teachings deal with neutronics and thermal-hydraulics for nuclear generated electricity. A firm foundation in thermodynamics and fluid mechanics in addition to hydrodynamics is a must.

Specialization in nuclear fusion includes electrodynamics and plasmas. This area is very much research oriented and training often terminates with a graduate level degree.

Specialization in nuclear medicine, includes courses dealing with doses and absorption of radiation in bodily tissues. Those who get competency in this area usually move into the medical field. Many nuclear engineers in this specialisation go on to become board licensed medical physicists or go to medical school and become a radiation oncologist. Research is also a common choice for graduates.

Naval Nuclear Power School

The U.S. Navy runs a program called Naval Nuclear Power School to train both officers and enlisted sailors for nuclear plant operation. While some officers have undergraduate backgrounds in nuclear engineering, those with degrees in physics, mathematics, or other engineering disciples are also accepted, whereas most of the enlisted students hold no college degrees at all. Despite this, they are prepared, through a rigorous training program (lasting between 65 weeks for Machinist's Mates and eighteen months for Electronics Technicians and Electrician's Mates), to operate the nuclear and steam plants aboard the navy's submarines and aircraft carriers. This training carries Department of Energy certification, and many sailors choose to work at civilian power plants after their six-year obligations are completed.

Professional Areas

Nuclear Fission

The United States gets about 20% of its electricity from nuclear power. This is a massive industry and keeping the supply of nuclear engineers plentiful will ensure its stability. Nuclear engineers in this field generally work, directly or indirectly, in the nuclear power industry or for government labs. Current research in industry is directed at producing economical, proliferation resistant reactor designs with passive safety features. Although government labs research the same areas as industry, they also study a myriad of other issues such as: nuclear fuels and nuclear fuel cycles, advanced reactor designs, and nuclear weapon design and maintenance. A principal pipeline for trained personnel for US reactor facilities is the Navy Nuclear Power Program.

Nuclear Fusion and Plasma Physics

Research areas include high-temperature, radiation-resistant materials, and plasma dynamics. Internationally, research is currently directed at building a prototype tokamak called ITER. The research at ITER will primarily focus on instabilities and diverter design refinement. Researchers in the USA are also building an inertial confinement experiment called the National Ignition Facility or NIF. NIF will be used to refine neutron transport calculations for the US stockpile stewardship initiative.

Nuclear Medicine and Medical Physics

An important field is nuclear medicine. From x-ray machines to MRI to PET,among many others, nuclear medicine provides most of modern medicine's diagnostic capability along with providing many treatment options.

Nuclear Materials and Nuclear Fuels

Nuclear materials research focuses on two main subject areas, nuclear fuels and irradiation-induced modification of materials. Improvement of nuclear fuels is crucial for obtaining increased efficiency from nuclear reactors. Irradiation effects studies have many purposes, from studying structural changes to reactor components to studying nano-modification of metals and semiconductors using ion-beams or particle accelerators.

Radiation Measurements and Imaging

Nuclear engineers and radiological scientists are interested in the development of more advanced ionizing radiation measurement and detection systems, and using these to improve imaging technologies. This includes detector design, fabrication and analysis, measurements of fundamental atomic and nuclear parameters, and radiation imaging systems, among other things.

Nuclear Engineering Organizations

- American Nuclear Society
- International Atomic Energy Agency
- Institution of Nuclear Engineers (London)

List of Colleges in India Offering Nuclear Engineering Degrees

College	Department (external links)	Degrees offered
Indian Institute of Technology, Kanpur	Nuclear Engg and Technology	M.Tech, PhD

List of U.S. Colleges Offering Nuclear Engineering degrees

College	*Department (external links)*	*Degrees offered*
Air Force Institute of Technology	Engineering Physics	MS, PhD
Kansas State University	Mechanical and Nuclear Engineering	BS, MS, PhD
Georgia Institute of Technology	Nuclear and Radiological Engineering	BS, MS, PhD
Idaho State University	[1]	BS
Massachusetts Institute of Technology	Nuclear Science and Engineering	BS, MS, PhD
North Carolina State University	Nuclear Engineering	BS, MS, PhD
Ohio State University	Nuclear Engineering	BS, MS, PhD
Oregon State University	Nuclear Engineering and Radiation Health Physics	BS, MS, PhD
Pennsylvania State University	Mechanical and Nuclear Engineering	BS, MS, PhD
Pennsylvania State University	Distance Learning Program in Nuclear Engineering	Meng
Purdue University	Nuclear Engineering	BS, MS, PhD
Rensselaer Polytechnic Institute	Mechanical, Aerospace & Nuclear Engineering	BS, MS, PhD

Texas A&M University	Nuclear Engineering	BS, MS, PhD
University of California, Berkeley	Nuclear Engineering	BS, MS, PhD
University of Cincinnati	Mechanical, Industrial and Nuclear Engineering	BS, MS, PhD
University of Florida	Nuclear and Radiological Engineering	BS, MS, PhD
University of Illinois at Urbana-Champaign	Nuclear, Plasma and Radiological Engineering	BS, MS, PhD
University of Maryland, College Park	Materials and Nuclear Engineering	MS, PhD
University of Massachusetts Lowell	Chemical Engineering	BS
University of Michigan Ann Arbor	Nuclear Engineering and Radiological Sciences	BS, MS, PhD
University of Missouri–Columbia	Nuclear Science and Engineering Institute	MS, PhD
University of Missouri–Rolla	Nuclear Engineering	BS, MS, PhD
University of New Mexico	Chemical and Nuclear Engineering	BS, MS, PhD
University of Tennessee at Knoxville	Nuclear Engineering	BS, MS, PhD
University of Texas at Austin	Nuclear and Radiation Engineering	MS, PhD
University of Wisconsin-Madison	Engineering Physics	BS, MS, PhD
United States Naval Academy	Nuclear Engineering	BS
United States Military Academy	Nuclear Engineering	BS

19

Petroleum Engineering

Petroleum Engineering is involved in the exploration and production activities of petroleum as an upstream end of the energy sector. Upstream, refers to the source of the petroleum, the petroleum deposit, usually buried deep beneath the earth's surface supplying flow to consumers as a river supplies the ocean. The diverse topics covered by petroleum engineering are closely related to the earth sciences. Petroleum engineering topics include economics, geology, geochemistry, geomechanics, geophysics, oil drilling, geopolitics, knowledge management, seismology, team building, team work, tectonics, thermodynamics, well logging, well completion, oil and gas production, reservoir development, and pipelining.

It is an increasingly technical profession that involves procuring reserves from places that predecessors deemed too difficult or not economic with the technology of the day or commodity prices. The use of high technology equipment, high speed computers, innovative materials, team management philosophies, statistics, probability analysis, and knowledge management, is usually coupled with the reality of only indirect measurement of most essential facts due to being buried under miles of earth. As such engineers must sometimes develop new techniques from one source to another, and one look at the number of patents held for use in the industry is testimony to the highly technical nature of this field.

As mistakes may be measured in millions of dollars, petroleum engineers are held to a high standard. Deepwater operations can arguably be compared to space travel in terms of technical challenges. Arctic conditions and conditions of extreme heat have to be contended with. High Temperature and High Pressure (HTHP) environments that have become increasingly commonplace in today's operations require the petroleum engineer to be savvy in topics as wide ranging as thermohydraulics, geomechanics, and intelligent systems.

Petroleum engineers must implement high technology plans with the use of manpower, highly coordinated and often in dangerous conditions. The drilling rig crew and machines they use become the remote partner of the petroleum engineer in implementing every drilling program. Understanding and accounting for the issues and communication challenges of building these teams remain just as vital to the petroleum engineer as ever.

The Society of Petroleum Engineers is the largest professional society for petroleum engineers and is a good source of information. Petroleum engineering education is available at dozens of universities in the United States and throughout the world - primarily in oil producing states — but not only top producers.

Petroleum engineers have historically been one of the highest paid engineering disciplines; this is offset by a tendency for mass layoffs when oil prices decline. Petroleum engineering offers a challenging blend of earth sciences, geology, operations, politics, advanced mathematics and the opportunity to risk massive amounts of money. The rewards for successful engineers range from high paying jobs to the opportunities to start oil companies.

Petroleum engineers divide themselves into several types:

- Reservoir engineers work to optimize production of oil and gas via proper well placement, production levels, and enhanced oil recovery techniques.
- Drilling engineers manage the technical aspects of drilling both production and injection wells.
- Subsurface engineers (also known as completion engineers) manage the interface between the reservoir and the well, including perforations, sand control, artificial lift, downhole flow control, and downhole monitoring equipment.

Major Petroleum Engineering Schools in World

- AGH University of Science and Technology
- Augusta Taurinorum
- Bangladesh University of Engineering &Technology
- Cairo U.
- California Polytechnic State U. - Pomona
- California Polytechnic State U. - San Luis Obispo
- California State U.
- Chulalongkorn U.
- College of the North Atlantic
- Colorado School of Mines
- Curtin University of Technology
- Dalhousie University
- Delft University of Technology
- Ecole Nationale Superieure du Petrole et des Moteurs (IFP School)
- Federal U. of Technology

- Freiberg University of Mining and Technology
- Heriot-Watt U.
- Imperial College of Science & Technology, Royal School of Mines
- Indian School of Mines
- Inst. Politecnico Nacional
- Inst. Teknologi Bandung
- Inst. of Management & Technology
- Instituto Technologico De Buenos Aires
- Istanbul Technical U.
- Kazakh National Technology U.
- Khazar U.
- King Fahd University of Petroleum & Minerals
- King Saud University
- Kuwait University
- Louisiana State University
- Maharashtra Institute of Technology
- Marietta College
- Mehran U. of Engineering & Tech
- Memorial University of Newfoundland
- Middle East Technical U.
- Montana Tech of the University of Montana
- National Development U.
- New Mexico Tech
- Northern Alberta Institute of Technology
- Norwegian University of Science & Technology
- Palawan State U.

- Penn State University
- Petroleum Training Inst.
- Rivers State University of Science & Technology
- Robert Gordon University
- South Dakota School of Mines
- Southern Alberta Institute of Technology
- Stanford University
- Stavanger College
- Sultan Qaboos University
- Technische Universitat Clausthal
- Texas A & M University
- Texas A & M University - Kingsville
- Texas Tech University
- Tomsk Polytechnic University
- Trisakti University
- Turkmenistan Polytechnic Institute
- United Arab Emirates University
- Universidad Autonoma Gabriel Rene Moreno
- Universidad Central de Venezuela
- Universidad Central del Ecuador
- Universidad Estadual do Norte Fluminense
- Universidad Industrial de Santander
- Universidad Nacional de Colombia
- Universidad Nacional de Cuyo
- Universidad Nacional de Mexico
- Universidad Nacional de Piura
- Universidad Nacional de la Patagonia

- Universidad Nacional del Comahue
- Universidad Nacionel de Ingenieria
- Universidad Surcolombiana
- Universidad de America
- Universidad de Oriente
- Universidad de Oriente Monagas
- Universidad del Zulia
- Universidade Estadual De Campinas
- Universitas Indonesia
- Universitatea Petrol - Gaze Din Ploiesti
- Universiti Teknologi Malaysia
- University of Aberdeen
- University of Alaska Fairbanks
- University of Alberta
- University of Benin
- University of Bologna
- University of Buenos Aires
- University of Calgary
- University of California Berkeley
- University of Engineering & Technology
- University of Houston
- University of Ibadan
- University of Kansas
- University of Leoben, Austria
- University of Louisiana at Lafayette
- University of Miskolc
- University of Missouri-Rolla

- University of New South Wales
- University of Oklahoma
- University of Pittsburgh
- University of Port Harcourt
- University of Saskatchewan
- University of Southern California
- University of Texas at Austin
- University of Tokyo
- University of Tulsa
- University of Uyo
- University of Western Australia
- University of Zagreb
- University of the West Indies
- UniversityU. of Karachi
- West Virginia University

20

Software Engineering

The term *software engineering* was popularized after 1968, during the 1968 NATO Software Engineering Conference (held in Garmisch, Germany) by its chairman F.L. Bauer, and has been in widespread use since.

Software engineering (SE) is the design, development, and documentation of software by applying technologies and practices from computer science, project management, engineering, application domains, interface design, digital asset management and other fields.

The term *software engineering* has been commonly used with a variety of distinct meanings:

- As the informal contemporary term for the broad range of activities that was formerly called programming and systems analysis;
- As the broad term for all aspects of the *practice* of computer programming, as opposed to the *theory* of computer programming, which is called computer science;
- As the term embodying the *advocacy* of a specific approach to computer programming, one that urges that it be treated as an engineering discipline rather than an art or a craft, and advocates the codification of recommended practices in the form of *software engineering methodologies*.

- Software engineering is "(1) the application of a systematic, disciplined, quantifiable approach to the development, operation, and maintenance of software, that is, the application of engineering to software," and "(2) the study of approaches as in (1)." – IEEE Standard 610.12

Scope and Focus

Software engineering is concerned with the conception, development and verification of a software system. This discipline deals with identifying, defining, realizing and verifying the required characteristics of the resultant software. These software characteristics may include: functionality, reliability, maintainability, availability, testability, ease-of-use, portability, and other attributes. Software engineering addresses these characteristics by preparing design and technical specifications that, if implemented properly, will result in software that can be verified to meet these requirements.

Software engineering is also concerned with the characteristics of the software development process. In this regard, it deals with characteristics such as cost of development, duration of development, and risks in development of software.

Software is often found in products and situations where very high reliability is expected, even under demanding conditions, such as monitoring and controlling nuclear power plants, or keeping a modern airliner aloft Such applications contain millions of lines of code, making them comparable in complexity to the most complex modern machines. For example, a modern airliner has several million physical parts (and the space shuttle about ten million parts), while the software for such an airliner can run to 4 million lines of code.

Technologies and Practices

Software engineers advocate many different technologies and practices, with much disagreement. This debate has gone on for 60 years and may continue forever. Software engineers use a wide variety of technologies and practices. Practitioners use a wide variety of technologies: compilers, code repositories, word processors. Practitioners use a wide variety of practices to carry out and coordinate their efforts: pair programming, code reviews and daily stand up meetings.The goal of every software engineer should be to bring an idea out of a previous planned model, which should be transparent and well documented(Omeke Tochukwu Benjamin(2006))

In spite of the enormous economic growth and productivity gains enabled by software, persistent complaints about the quality of software remain.

Nature of SE?

Software engineering resembles many different fields in many different ways. The following paragraphs make some simple comparisons.

Mathematics

Programs have many mathematical properties. For example the correctness and complexity of many algorithms are mathematical concepts that can be rigorously proven. Programs are finite, so in principle, developers could know many things about a program in a rigorous mathematical way. The use of mathematics within software engineering is often called formal methods. However, computability theory shows that not everything useful about a program can be proven. Mathematics works best for small pieces of code and has difficulty scaling up.

Engineering

Software Engineering is considered by many to be an engineering discipline because there are pragmatic approaches and expected characteristics of engineers. Proper analysis, documentation, and commented code are signs of an engineer. David Parnas has argued that software engineering is engineering. Programs have many properties that can be measured. For example, the performance and scalability of programs under various workloads can be measured. The effectiveness of caches, bigger processors, faster networks, newer databases are engineering issues. Mathematical equations can sometimes be deduced from the measurements. Mathematical approaches work best for system-wide analysis, but often are meaningless when comparing different small fragments of code.

Manufacturing

Programs are built in as a sequence of steps. By properly defining and carrying out those steps, much like a manufacturing assembly line, advocates hope to improve the productivity of developers and the quality of final programs. This approach inspires the many different processes and methodologies. While others, such as the authors of the Programmer's Stone, contend this view "[is] in fact claiming to be able to implement an Artificial Intelligence that simulates a production line designer".

Project Management

Commercial (and many non-commercial) software projects require management. There are budgets and schedules to set. People to hire and lead. Resources (office space, computers) to acquire. All of this fits more appropriately within the purview of management.

Audio and Visual Art

Programs contain many artistic elements, akin to writing or painting. User interfaces should be aesthetically pleasing and provide optimal audio and visual communication to end-users. What is considered "good design" is often subjective, and may be decided by one's own sense of aesthetics. Because graphic artists create graphic elements for graphical user interfaces, graphic design often overlaps interaction design in the position of an interface designer.

User interfaces may require technical understanding including graphical integration with code, computer animation technology, automation of graphic production, integrating graphics with sound editing technology, and mathematical application. One could say that "audiovisual engineering" is required.

User interfaces with user-read text and voice may also be enhanced from professional copywriting and technical writing.

Code should be aesthetically pleasing to programmers. Even the decision of whether a variable name or class name is clear and simple is an artistic question. Donald Knuth asserted that programming is an art.

Performance

The act of writing software requires that developers summon the energy to find the answers they need while they are at the keyboard. Creating software is a performance that resembles what athletes do on the field, and actors and musicians do on stage. Some argue that SEs need inspiration to spark the creation of code. Sometimes a creative spark is needed to create the architecture or to develop a unit of code to solve a particularly intractable problem. Others argue that discipline is the key attribute. Pair programming emphasizes this point of view. Both

Kent Beck and Watts Humphrey have argued this emphasis. See also Performance Engineering.

Is SE (or should SE be) a branch of programming, a branch of computer science, a branch of traditional engineering, or a field that stands on its own? There is considerable debate over this. This has important implications for professionalism, licensing, and ethics. Licensing is a polarizing issue: some fiercely advocate it while others staunchly oppose it.

Branch of Programming

Programming emphasizes writing code, independent of projects and customers. Software engineering emphasizes writing code in the context of projects and customers by making plans and delivering applications. As a branch of programming, SE would probably have no significant licensing or professionalism issues.

Branch of Computer Science

Many believe that software engineering is a part of computer science, because of their close historical connections and their relationship to mathematics. They advocate keeping SE a part of computer science. Both computer science and software engineering care about programs. Computer science emphasizes the theoretical, eternal truths while software engineering emphasizes practical, everyday usefulness. Some argue that computer science is to software engineering as physics and chemistry are to traditional engineering. As a branch of computer science, SE would probably have few licensing or professionalism concerns.

Branch of Engineering

Some SE academics and practitioners, such as David Parnas and Steve McConnell, have advocated treating SE an engineering discipline. Advocates for this view argue that the practice of engineering involves the use of mathematics, science, and the technology of the day, to build trustworthy products that are "fit for purpose", a description that applies as well to SE as to any other engineering discipline. As a branch of engineering, SE would probably adopt the engineering model of licensing and professionalism.

Freestanding Field

Recently, software engineering has been finding its own identity and emerging as an important freestanding field. Practitioners are slowly realizing that they form a huge community in their own right. Software engineering may need to create a form of regulation/licensing appropriate to its own circumstances. It is arguable that licensing (in the United States) is inappropriate because the creation of software represents a form of writing, and requiring people to be licensed in order to write computer programs may be a violation of the First Amendment. Requiring software engineers to be licensed by a government bureaucracy would make persons who create software without a license into criminals, even if they give their software away, same as practicing medicine or law without a license, even for free, is a criminal offense.

The U.S. Bureau of Labor Statistics classifies computer software engineers as a subcategory of "computer specialists", along with occupations such as computer scientist, programmer, and network administrator. The BLS classifies all other engineering disciplines, including computer hardware engineers, as "engineers".

The Software Engineering Profession

As of 2004, the U. S. Bureau of Labor Statistics counts 760,840 software engineers holding jobs in the U.S.; for comparison, in the U.S. there are some 1.4 million practitioners employed in all other engineering disciplines combined. There are estimated to be about 1.5 million practitioners in the E.U., Asia, and elsewhere. SE pioneers include Barry Boehm, Fred Brooks, C. A. R. Hoare, and David Parnas. There is extensive debate about what SE is, who qualifies as a SE, who sets the standards, etc.

The use of "software engineer" as a corporate job title dates from at least 1990. The person who has the job title "software engineer" and performs his job according to the job description for a "software engineer" is generally thought of as a software engineer.

However, Bachelor's degrees in Software Engineering have become available from many well known universities. People who are from outside the corporate setting may believe that there are currently no widely accepted criteria for distinguishing someone who is a software engineer from someone who is not a software engineer. In addition, the industry is in the midst of a complex debate on the licensing of practicing software engineers. The two sides of the debate consider whether the job is a creative art or a repeatable science. In actual practice, the existing system of checking the work of the 'software engineer' when required by law by a state licensed professional engineer is considered adequate for legal liability reasons.

Some people believe that software development *is a more appropriate term than* software engineering *for the process of creating software. Pete McBreen, (author of "Software Craftsmanship: The New Imperative" (ISBN 0-201-73386-2)), argues that the term Software Engineering implies levels of rigor and proven processes that are not appropriate for all types of software development. He argues strongly for 'craftsmanship' as a more appropriate metaphor because that term brings into sharper focus the skills of the developer as the key to success instead of the "manufacturing" process. Using a more traditional comparison, just as not everyone who works in construction is a civil engineer, not everyone who can write code is a software engineer.*

Some people dispute the notion that the field is mature enough to warrant the title "engineering". Opposition also comes from the traditional engineering disciplines, whose practitioners usually object to the use of the title "engineer" by anyone who has not gone through an accredited program of engineering education. In each of the last few decades, at least one radical new approach has entered the mainstream of software development (e.g. Structured Programming, Object Orientation, ...), implying that the field is still changing too rapidly to be considered an engineering discipline. Other people would argue that the supposedly radical new approaches are actually evolutionary rather than revolutionary, the mere introduction of new tools rather than fundamental changes.

Education

People from many different educational backgrounds make important contributions to SE. Today, software engineers earn software engineering, computer engineering or computer science degrees.

Software Degrees

About half of all practitioners today have computer science degrees. A small, but growing, number of practitioners have software engineering degrees. In 1996, Rochester Institute of Technology established the first BSSE degree program in the United States but was beaten to ABET accreditation by Milwaukee School of Engineering. Both programs received ABET accreditation in 2003. Since then, software engineering undergraduate degrees have been established at many universities. A standard international curriculum for undergraduate software engineering degrees was recently defined by the CCSE. As of 2004, in the U.S., about 50 universities offer software engineering degrees, which teach both computer science and engineering principles and practices. The first graduate software engineering degree (MSSE) was established at Seattle University in 1979. Since then graduate software engineering degrees have been made available from many more universities.

Domain Degrees

Some practitioners have degrees in application domains, bringing important domain knowledge and experience to projects. In MIS, some practitioners have business degrees. In embedded systems, some practitioners have electrical or computer engineering degrees, because embedded software often requires a detailed understanding of hardware. In medical software, some practitioners have medical informatics, general medical, or biology degrees.

Other Degrees

Some practitioners have mathematics, science, engineering, or other technical degrees. Some have philosophy, or other non-technical degrees. And, some have no degrees. Note that Barry Boehm earned degrees in mathematics and Edsger Dijkstra earned degrees in physics.

Employment Opportunities

Most software engineers work as employees or contractors. Software engineers work with businesses, government agencies (civilian or military), and non-profit organizations. Some software engineers work for themselves as freelancers. Some organizations have specialists to perform each of the tasks in the software development process. Other organizations required software engineers to do many or all of them. In large projects, people may specialize in only one role. In small projects, people may fill several or all roles at the same time. Specializations include: in industry (analysts, architects, developers, testers, technical support, managers) and in academia (educators, researchers).

There is considerable debate over the future employment prospects for Software Engineers and other IT Professionals. For example, an online futures market called the Future of IT Jobs in America attempts to answer whether there will be more IT jobs, including software engineers, in 2012 than there were in 2002.

Certification

Certification of software engineers is a contentious issue. Some see it as a tool to improve professional practice.

Most successful certification programs in the software industry are oriented toward specific technologies, and are managed by the vendors of these technologies. These certification programs are tailored to the institutions that would employ people who use these technologies. General certification of software practitioners has struggled. The ACM had a professional certification program in the early 1980s, which was discontinued due to lack of interest. Today, the IEEE is certifying software professionals, but only about 500 people have passed the exam by March 2005. In Canada the use of the title of Software Engineer is regulated, and Information Systems Professional certification is used.

For the localities that do not license or certify software engineers, some hiring classifications are made based on education and experience. Classification levels may include: entry-level, mid-level, and senior. Typical entry-level software engineers have a bachelor's degree and zero to five years of experience. Typical mid-level software engineers have a bachelor's or master's degree and have five to ten years of experience. Typical senior-level software engineers have an advanced degree and have ten or more years of experience. Note that these are only guidelines that are trends seen in hiring practices and that many exceptions exist.

Many students in the developed world have avoided degrees related to software engineering because of the fear of offshore outsourcing (importing software products or services from other countries) and of being displaced by foreign visa workers. Although government statistics do not currently show a threat to software engineering itself; a related career, computer programming does appear to have been affected. Often one is expected to start out as a computer programmer before being promoted to software engineer. Thus, the career path to software engineering may be rough, especially during recessions.

Some career counselors suggest a student also focus on "people skills" and business skills rather than purely technical skills because such "soft skills" are allegedly more difficult to offshore. It is the quasi-management aspects of software engineering that appear to be what has kept it from being impacted by globalization.

Many fields are closely related to software engineering; here are some key similarities and distinctions. Comparing SE with other fields helps explain what SE is and helps define what SE might or should become. There is considerable debate over which fields SE most resembles (or should most resemble). These complex and inexact comparisons explain why some see software engineering as its own field.

Software engineering has a long evolving history. Both the tools that are used and the applications that are written have evolved over time. It seems likely that software engineering will continue evolving for many decades to come.

60 Year Time Line

- 1940s: First computer users wrote machine code by hand.
- 1950s: Early tools, such as macro assemblers and interpreters were created and widely used to improve productivity and quality. First generation optimizing compilers.
- 1960s: Second generation tools like optimizing compilers and inspections were being used to improve productivity and quality. The concept of software engineering was widely discussed. First really big (1000 programmer) projects. Commercial mainframes and custom software for big business. The influential 1968 NATO Conference on Software Engineering was held.
- 1970s: Collaborative software tools, such as Unix, code repositories, make, and so on. Minicomputers and the rise of small business software.
- 1980s: Personal computers and personal workstations become common. Commensurate rise of consumer software.
- 1990s: Object-oriented programming and agile processes like Extreme programming gain mainstream acceptance. Computer memory capacity sky rockets and prices drop drastically. These new technology allows software to grow more complex. The WWW and hand-held computers make software even more widely available.
- 2000s: Managed code and interpreted platforms such as .NET, PHP, Python and Java make writing

software easier than ever before. Offshore outsourcing changes the nature and focus of software engineering careers.

Current Trends

Software engineering is a young discipline, and is still developing. The directions in which software engineering is developing include:

Aspects

Aspects help programmers deal with *-ilities* by providing tools to add or remove boilerplate code from many areas in the source code. Aspects describe how all objects or functions should behave in particular circumstances. For example, aspects can add debugging, logging, or locking control into all objects of particular types. Researchers are currently working to understand how to use aspects to design general-purpose code. Related concepts include generative programming and templates.

Agile

Agile software development guides software development projects that evolve rapidly with changing expectations and competitive markets. Proponents of agile software development believe that heavy, document-driven processes (like TickIT, CMM and ISO 9000) are fading in importance. Some people believe that companies and agencies export many of the jobs that can be guided by heavy-weight processes. Related concepts include extreme programming and lean software development.

Experimental

Experimental software engineering is a branch of software engineering interested in devising experiments on software, in collecting data from these experiments, and in devising laws and theories from this data. Proponents of experimental software engineering advocate that the nature of software is such that we can advance the knowledge on software through experiments only.

Software Product Lines

Software Product Lines is a systematic way of producing *families* of software systems, instead of creating a succession of completely individual products. The Software Product Lines approach is an attempt to industrialize the software development process.

The *Future of Software Engineering* conference (FOSE), held at ICSE 2000, documented the state of the art of SE in 2000 and listed many problems to be solved over the next decade. The Feyerabend project attempts to discover the future of software engineering by seeking and publishing innovative ideas.

In 2006, Money Magazine and Salary.com rated software engineering as the best job in America in terms of growth, pay, stress levels, flexibility in hours and working environment, creativity, and how easy it is to enter and advance in the field.

Several academic conferences devoted to software engineering are held every year. There are also many other academic conferences every year devoted to special topics within SE, such as programming languages, requirements, testing, and so on.

ICSE

The biggest and oldest conference devoted to software engineering is the International Conference on Software Engineering. This conference meets every year to discuss improvements in research, education, and practice.

COMPSAC

The Annual International Computer Software and Applications Conference was first held in Chicago in 1977 and is designated as the IEEE Computer Society signature conference on software technology and applications.

ESEC

The European Software Engineering Conference.

FSE

The Foundations of Software Engineering conference is held every year, alternating between Europe and North America. It emphasizes theoretical and foundational issues.

CUSEC

Conferences dedicated to inform undergraduate students like the annual Canadian University Software Engineering Conference are also very promising for the future generation. It is completely organized by undergraduate students and lets different Canadian universities interested in Software Engineering host the conference each year. Past guests include Kent Beck, Joel Spolsky, Philippe Kruchten, Hal Helms, Craig Larman, David Parnas as well as university professors and students.

SEPG

The annual Software Engineering Process Group conference, sponsored by the Carnegie Mellon Software Engineering Institute (SEI), is a conference and exhibit showcase for systems and software engineering professionals. The four-day event emphasizes systematic improvement of people, processes, and technology.

INFORMATICS-INFORMATIQUE

The annual Canadian information technology, data

processing and software engineering symposium, sponsored by the Canadian Information Processing Society. First held in 1958.

ICALEPS

International Conference on Accelerator and Large Experimental Physics Control Systems Conference. Biennial conference covering software engineering for large scale scientific control systems. First held in 1987.

APSEC

Asia Pacific Software Engineering Conference.

UYMS

National Software Engineering Symposium (in Turkish: Ulusal Yazilim Muhendisligi Sempozyumu) (not available in English). Biennial symposium first held in Ýzmir, Turkey in 2003.

Organizations

- Association for Computing Machinery (ACM)
- British Computer Society (BCS)
- Australian Computer Society (ACS)
- Canadian Information Processing Society (CIPS) - Information Systems Professional certification.
- IEEE Computer Society
- Lero, the Irish Software Engineering Research Centre
- RUSSOFT Association
- Software Engineering Institute (SEI)
- Software Association of Oregon (SAO)

REFERENCES

Akram I. Salah (2002-04-05). Engineering an Academic Program in Software Engineering. 35th Annual Midwest Instruction and Computing Symposium.

Carter, A.G.; C. Sanger (1997). Thinking About Thinking. *The Programmers' Stone*. Alan Carter.

David Budgen, Pearl Brereton, Barbara Kitchenham, Stephen Linkman (2004-12-14). Realizing Evidence-based Software Engineering.

Grübler, Arnulf (2003). Technology and Global Change. Cambridge University Press.

Kalwarski, Tara; Daphne Mosher, Janet Paskin and Donna Rosato (2006). Best Jobs in America. *MONEY Magazine*. CNN.

McConnell, Steve *(August 2003). Professional Software Development: Shorter Schedules, Better Projects, Superior Products, Enhanced Careers. Boston, MA: Addison-Wesley.*

Mills, Harlan D., J. R. Newman, and C. B. Engle, Jr., "An Undergraduate Curriculum in Software Engineering," in *Deimel, Lionel E. (1990). Software Engineering Education: SEI Conference 1990, Pittsburgh, Pennsylvania, USA, April 2-3,.... Springer.*

Parnas, David L. *(1998)*. "Software Engineering Programmes are not Computer Science Programmes". *Annals of Software Engineering 6: 19–37.*, p. 19.

Parnas, David L. *(1998)*. "Software Engineering Programmes are not Computer Science Programmes". *Annals of Software Engineering 6: 19–37.*, p. 20.

Pecht, Michael (1995). Product Reliability, Maintainability, and Supportability Handbook. CRC Press.

Pehrson, Ronald J. (January 1996). "Software Development for the Boeing 777". CrossTalk: The Journal of Defense Software Engineering.,

The 1968 NATO Software Engineering Conference

U.S Department of Labor and Statistics *The 2000 Standard Occupational Classification (SOC) System: 15-0000 Computer and Mathematical Occupations*

21

Structural Engineering

Basically, Structural engineering is a field of engineering that deals with the design of any structural system(s), the purpose of which is to support and resist various loads.

Most commonly, structural engineers are involved in the design of buildings and nonbuilding structures, but also play an essential role in designing machinery where structural integrity of the design item is a matter of safety and reliability. Large man-made objects—everything from furniture to medical equipment and from vehicles (trucks, aircraft, spacecraft and watercraft) to cranes—require the input of a structural engineer.

In building construction, the structural engineering field is a subset of civil engineering. In a practical sense, structural engineering is largely the application of Newtonian mechanics to the design of structural elements and systems that support buildings, bridges, walls (including retaining walls), dams, tunnels, etc.

Structural engineers ensure that their designs satisfy a given design intent predicated on safety (i.e. structures do not collapse without due warning) and on serviceability (i.e. floor vibration and building sway do not result in discomfort for the occupants. Structural engineers are responsible for making efficient use of funds and materials to achieve these goals. Entry-level structural engineers may design simple beams, columns, and floors of a new building, including calculating the loads on each member and the load capacity of various building materials (steel, timber, masonry, concrete). More experienced engineers would render more difficult structures, for example considering the physics of moisture, heat and energy inside the building components.

The structural engineering field is often subdivided into bridge engineering and structural engineering for buildings. Additionally, structural engineers often further specialize into special structure manufacture or construction, such as pipeline engineering or industrial structures.

Structural loads on structures are generally classified as: live loads, such as the weight of occupants and furniture in a building, the forces of wind or weights of water, and the forces due to seismic activity such as an earthquake; and dead loads, including the weight of the structure itself and all major architectural components; and live roof loads, such as material and manpower loading the structure during construction. Limiting design criteria are often forces of nature like winds, earthquakes and tsunamis. In recent years, however, reinforcing structures against sabotage has taken on increased importance.

Major Institutions for Structural Engineering

- Auburn University (Alabama)
  www.auburn.edu
- California State University, Northridge
  www.csun.edu
- California State Polytechnic University, Pomona
  www.csupomona.edu
- California, Berkeley, University of
  www.berkeley.edu

- California, Irvine, University of
  www.uci.edu
- California, Los Angeles, University of
  www.ucla.edu
- California, San Diego, University of
  www.ucsd.edu
- Case Western Reserve University (Ohio)
  www.cwru.edu
- Catholic University of America (District of Columbia)
  www.cua.edu
- Central Florida, University of
  www.ucf.edu
- Clemson University (South Carolina)
  www.clemson.edu
- Cleveland State University (Ohio)
  www.csuohio.edu
- Colorado at Boulder, University of
  www.colorado.edu
- Colorado State University
  www.colostate.edu
- Cornell University (New York)
  www.cornell.edu
- Dayton, University of (Ohio)
  www.udayton.edu
- Delaware, University of
  www.udel.edu
- Florida International University
  www.fiu.edu
- Houston, University of (Texas)
  www.uh.edu
- Howard University (District of Columbia)
  www.howard.edu
- Illinois Institute of Technology
  www.iit.edu
- Iowa State University of Science and Technology
  www.iastate.edu

- Lehigh University (Pennsylvania)
  www.lehigh.edu
- Louisiana State University and Agricultural and Mechanical College
  www.lsu.edu
- Maine (Orono), University of
  www.umaine.edu
- Marquette University (Wisconsin)
  www.mu.edu
- Massachusetts Institute of Technology
  www.mit.edu
- Memphis, University of (Tennessee)
  www.memphis.edu
- Michigan Technological University
  www.mtu.edu
- Milwaukee School of Engineering (Wisconsin)
  www.msoe.edu
- Mississippi State University
  www.msstate.edu
- Missouri-Columbia, University of
  www.missouri.edu
- Missouri-Rolla, University of
  www.umr.edu
- New York at Buffalo, State University of
  www.buffalo.edu
- North Dakota, University of
  www.und.nodak.edu
- Northwestern University
  www.northwestern.edu/
- Ohio University
  www.ohio.edu/
- Oklahoma, University of
  www.ou.edu
- Pennsylvania State University, University Park Campus
  www.psu.edu

- Princeton University (New Jersey)
  www.princeton.edu
- Rensselaer Polytechnic Institute (New York)
  www.rpi.edu
- Rhode Island, University of
  www.uri.edu
- Southern California, University of
  www.usc.edu
- Stevens Institute of Technology (New Jersey)
  www.stevens.edu
- Texas A&M University, College Station
  www.tamu.edu
- Texas at Austin, University of
  www.engr.utexas.edu/
- Toledo, University of (Ohio)
  www.utoledo.edu
- Tufts University (Massachusetts)
  www.tufts.edu
- Virginia, University of
  www.virginia.edu
- Washington University in St. Louis (Missouri)
  www.wustl.edu
- Washington, University of
  www.washington.edu
- Western Michigan University
  www.wmich.edu
- Worcester Polytechnic Institute (Massachusetts)
  www.wpi.edu

22

Tissue Engineering

Tissue engineering is the use of a combination of cells, engineering materials, and suitable biochemical factors to improve or replace biological functions. Probably the first definition of tissue engineering was by Langer and Vacanti who stated it to be "an interdisciplinary field that applies the principles of engineering and life sciences toward the development of biological substitutes that restore, maintain, or improve tissue function or a whole organ". MacArthur and Oreffo (as cited in "References") defined tissue engineering as "understanding the principles of tissue growth, and applying this to produce functional replacement tissue for clinical use." A further description goes on to say that an "underlying supposition of tissue engineering is that the employment of natural biology of the system will allow for greater success in developing therapeutic strategies aimed at the replacement, repair, maintenance, and/or enhancement of tissue function." These more general definitions are driven in part by recent scientific progress with completely autologous approaches. That is, many groups (Nicolas L'Heureux at Cytograft Tissue Engineering, Julie Campbell at University of Queensland etc, Loex laboratories at the Universite of Laval etc.) are demonstrating functional tissue engineered devices/organs without using synthetic biomaterials/scaffolds. These recent approaches are clearly based more on an understanding of cell biology than materials science.

In 2003, the NSF published a report titled "The Emergence of Tissue Engineering as a Research Field", which gives a thorough description of the history of this field.

While the semi-official definition of tissue engineering covers a broad range of applications, in practice the term has come to represent applications that repair or replace structural tissues (i.e., bone, cartilage, blood vessels, bladder, etc...). These are tissues that function by virtue of their mechanical properties. A closely related (and older) field is cell transplantation. This field is concerned with the transplantation of cells that perform a specific biochemical function (e.g., an artificial pancreas, or an artificial liver). The term regenerative medicine is often used synonymously with tissue engineering, although those involved in regenerative medicine place more emphasis on the use of stem cells to produce tissues.

Cells

Tissue engineering solves problems by using living cells as engineering materials. These could be artificial skin that includes living fibroblasts, cartilage repaired with living chondrocytes, or other types of cells used in other ways.

Cells became available as engineering materials when scientists at Geron Corp. discovered how to extend telomeres in 1998, producing an immortalized cell line. Before this, laboratory cultures of healthy, noncancerous mammalian cells would only divide a fixed number of times, up to the Hayflick limit.

From fluid tissues such as blood, cells are extracted by bulk methods, usually centrifugation or apheresis.

From solid tissues, extraction is more difficult. Usually

the tissue is minced, and then digested with the enzymes trypsin or collagenase to remove the extracellular matrix that holds the cells. After that, the cells are free floating, and extracted using centrifugation or aspheresis.

Digestion with trypsin is very dependent on temperature. Higher temperatures digest the matrix faster, but create more damage. Collagenase is less temperature dependent, and damages fewer cells, but takes longer and is a more expensive reagent.

Cells are often categorized by their source:

- *Autologous* cells are obtained from the same individual to which they will be reimplanted. Autologous cells have the fewest problems with rejection and pathogen transmission, however in some cases might not be available. For example in genetic disease suitable autologous cells are not available. Also very ill or elderly persons, as well as patients suffering from severe burns, may not have sufficient quantities of autologous cells to establish useful cell lines. Moreover since this category of cells needs to be harvested from the patient, there are also some concerns related to the necessity of performing such surgical operations that might lead to donor site infection or chronic pain. Autologous cells also must be cultured from samples before they can be used: this takes time, so autologous solutions may not be very quick. Recently there as been a trend towards the use of mesenchymal stem cells from bone marrow and fat. These cells can differentiate into a variety of tissue types, including bone, cartilage, fat, and nerve. A large number of cells can be easily and quickly isolated from fat, thus opening the potential for large numbers of cells to be quickly and easily obtained. Several companies have been founded to capitalize on this technology, the most successful at this time being Cytori Therapeutics.

- *Allogenic* cells come from the body of a donor of the same species. While there are some ethical constraints to the use of human cells for *in vitro* studies, the employment of dermal fibroblasts from human foreskin has been demonstrated to be immunologically safe and thus a viable choice for tissue engineering of skin.
- *Xenogenic* cells are those isolated from individuals of another species. In particular animal cells have been used quite extensively in experiments aimed at the construction of cardiovascular implants.
- *Syngeneic' or isogenic* cells are isolated from genetically identical organisms, such as twins, clones, or highly inbred research animal models.
- *Primary* cells are from an organism.
- *Secondary* cells are from a cell bank.
- *Stem cells* are undifferentiated cells with the ability to divide in culture and give rise to different forms of specialized cells. According to their source stem cells are divided into "adult" and "embryonic" stem cells, the first class being multipotent and the latter mostly pluripotent; some cells are totipotent, in the earliest stages of the embryo. While there is still a large ethical debate related with the use of embryonic stem cells, it is thought that stem cells may be useful for the repair of diseased or damaged tissues, or may be used to grow new organs.

Engineering Materials

Cells as found above are generally implanted or 'seeded' into an artificial structure capable of supporting three-dimensional tissue formation. These scaffolds are often critical, both ex vivo as well as in vivo, to recapitulating the in vivo milieu and allowing cells to influence their own microenvironments. Such devices, usually referred to as scaffolds, serve at least one of the following purposes:

- Allow cell attachment and migration
- Deliver and retain cells and biochemical factors
- Enable diffusion of vital cell nutrients and expressed products
- Exert certain mechanical and biological influences to modify the behaviour of the cell phase

To achieve the goal of tissue reconstruction, scaffolds must meet some specific requirements. A high porosity and an adequate pore size are necessary to facilitate cell seeding and diffusion throughout the whole structure of both cells and nutrients. Biodegradability is essential since scaffolds need to be absorbed by the surrounding tissues without the necessity of a surgical removal. The rate at which degradation occurs has to coincide as much as possible with the rate of tissue formation: this means that while cells are fabricating their own natural matrix structure around themselves, the scaffold is able to provide structural integrity within the body and eventually it will break down leaving the neotissue, newly formed tissue which will take over the mechanical load. Injectability is also important for clinical uses.

Many different materials (natural and synthetic, biodegradable and permanent) have been investigated. Most of these materials have been known in the medical field before the advent of tissue engineering as a research topic, being already employed as bioresorbable sutures. Examples of these materials are collagen or some linear aliphatic polyesters.

New biomaterials have been engineered to have ideal properties and functional customization: injectability, synthetic manufacture, biocompatability, non-immunogenicity, transparency, nano-scale fibers, low concentration, resorption rates, etc. PuraMatrix, originating from the MIT labs of Zhang, Rich, Grodzinsky and Langer is one of these new biomimetic scaffold families which has now been commercialized and is impacting clinical tissue engineering. A commonly used synthetic material is PLA - polylactic acid. This is a polyester which degrades within the human body to form lactic acid, a naturally occurring chemical which is easily removed from the body. Similar materials are polyglycolic acid (PGA) and polycaprolactone (PCL): their degradation mechanism is similar to that of PLA, but they exhibit respectively a faster and a slower rate of degradation compared to PLA.

Scaffolds may also be constructed from natural materials: in particular different derivatives of the extracellular matrix have been studied to evaluate their ability to support cell growth. Proteic materials, such as collagen or fibrin, and polysaccharidic materials, like chitosan or glycosaminoglycans (GAGs), have all proved suitable in terms of cell compatibility, but some issues with potential immunogenicity still remains. Among GAGs hyaluronic acid, possibly in combination with cross linking agents (e.g. glutaraldehyde, water soluble carbodiimide, etc...), is one of the possible choices as scaffold material. Functionalized groups of scaffolds may be useful in the delivery of small molecules (drugs) to specific tissues.

Synthesis of Tissue Engineering Scaffolds

A number of different methods has been described in literature for preparing porous structures to be employed as tissue engineering scaffolds. Each of these techniques presents its own advantages, but none is devoid of drawbacks.

- *Nanofiber Self-Assembly:* Molecular self-assembly is one of the few methods to create biomaterials with properties similar in scale and chemistry to that of the natural in vivo extracellular matrix (ECM). Moreover, these hydrogel scaffolds have shown superior in vivo toxicology and biocompatibility compared with traditional macroscaffolds and animal-derived materials. PuraMatrix synthetic peptide hydrogels exemplify this category.
- *Textile technologies:* these techniques include all the approaches that have been successfully employed for the preparation of non-woven meshes of different polymers. In particular non-woven polyglycolide structures have been tested for tissue engineering applications: such fibrous structures have been found useful to grow different types of cells. The principal drawbacks are related to the difficulties of obtaining high porosity and regular pore size.
- *Solvent Casting & Particulate Leaching (SCPL):* this approach allows the preparation of porous structures with regular porosity, but with a limited thickness. First the polymer is dissolved into a suitable organic solvent (e.g. polylactic acid could be dissolved into dichloromethane), then the solution is cast into a mold filled with porogen particles. Such porogen can be an inorganic salt like sodium chloride, crystals of saccharose, gelatin spheres or paraffin spheres. The size of the porogen particles will affect the size of the scaffold pores, while the polymer to porogen ratio is directly correlated to the amount of porosity of the final structure. After the polymer solution has been cast the solvent is allowed to fully evaporate, then the composite structure in the mold is immersed in a bath of a liquid suitable for dissolving the porogen: water in case of sodium chloride, saccharose

and gelatin or an aliphatic solvent like hexane for paraffin. Once the porogen has been fully dissolved a porous structure is obtained. Other than the small thickness range that can be obtained, another drawback of SCPL lies in its use of organic solvents which must be fully removed to avoid any possible damage to the cells seeded on the scaffold.

- *Gas Foaming:* to overcome the necessity to use organic solvents and solid porogens a technique using gas as a porogen has been developed. First disc shaped structures made of the desired polymer are prepared by means of compression molding using a heated mold. The discs are then placed in a chamber where are exposed to high pressure CO_2 for several days. The pressure inside the chamber is gradually restored to atmospheric levels. During this procedure the pores are formed by the carbon dioxide molecules that abandon the polymer, resulting in a sponge like structure. The main problems related to such a technique are caused by the excessive heat used during compression molding (which prohibits the incorporation of any temperature labile material into the polymer matrix) and by the fact that the pores do not form an interconnected structure.
- *Emulsification/Freeze-drying:* this technique does not require the use of a solid porogen like SCPL. First a synthetic polymer is dissolved into a suitable solvent (e.g. polylactic acid in dichloromethane) then water is added to the polymeric solution and the two liquids are mixed in order to obtain an emulsion. Before the two phases can separate, the emulsion is cast into a mold and quickly frozen by means of immersion into liquid nitrogen. The frozen emulsion is subsequently freeze-dried to remove the dispersed water and the solvent, thus leaving a solidified,

porous polymeric structure. While emulsification and freeze-drying allows a faster preparation if compared to SCPL, since it does not require a time consuming leaching step, it still requires the use of solvents, moreover pore size is relatively small and porosity is often irregular. Freeze-drying by itself is also a commonly employed technique for the fabrication of scaffolds. In particular it is used to prepare collagen sponges: collagen is dissolved into acidic solutions of acetic acid or hydrochloric acid that are cast into a mold, frozen with liquid nitrogen then liophylized.

- *Liquid-liquid phase separation:* similar to the previous technique, this phase separation procedure requires the use of a solvent with a low melting point that is easy to sublime. For example dioxane could be used to dissolve polylactic acid, then phase separation is induced through the addition of a small quantity of water: a polymer-rich and a polymer-poor phase are formed. Following cooling below the solvent melting point and some days of vacuum-drying to sublime the solvent a porous scaffold is obtained. Liquid-liquid phase separation presents the same drawbacks of emulsification/freeze-drying.

- *CAD/CAM Technologies:* since most of the above described approaches are limited when it comes to the control of porosity and pore size, computer assisted design and manufacturing techniques have been introduced to tissue engineering. First a three-dimensional structure is designed using CAD software, then the scaffold is realized by using ink-jet printing of polymer powders or through Fused Deposition Modeling of a polymer melt.

Assembly Methods

One of the continuing, persistent problems with tissue engineering is mass transport limitations. Engineered tissues generally lack an initial blood supply, thus making it difficult for any implanted cells to obtain sufficient oxygen and nutrients to survive, and/or function properly.

Self-assembly may play an important role here, both from the perspective of encapsulating cells and proteins, as well as creating scaffolds on the right physical scale for engineered tissue constructs and cellular ingrowth.

It might be possible to print organs, or possibly entire organisms. A recent innovative method of construction uses an ink-jet mechanism to print precise layers of cells in a matrix of thermoreversable gel. Endothelial cells, the cells that line blood vessels, have been printed in a set of stacked rings. When incubated, these fused into a tube.

Dr. Anthony Atala of Wake Forest University has successfully implanted artificially grown bladders into seven out of approximately 20 human test subjects as part of a long-term experiment, with demonstrated positive benefits to the recipients thus far. Lab-grown tissue was successfully used to repair knee cartilage.

REFERENCES

Davis, M.E., et al., Injectable Self-Assembling Peptide Nanofibers Create Intramyocardial Microenvironments for Endothelial Cells. Circulation 111:442-450 (2005).

Holmes, T.C., et al., Extensive neurite outgrowth and active synapse formation on self-assembling peptide scaffolds. PNAS USA 97 :6728 (2000).

Langer, R & Vacanti JP, Tissue engineering. Science 260, 920-6; 1993 (cited more than 1000 times!)

Ma, Peter X.: Scaffolds for tissue fabrication - *Materials Today*, May 2004, 30-40.

MacArthur, B. D. & Oreffo, R.O.C. (2005). "*Bridging the gap*". Nature 433, 19.

Mikos, A. G. & Temenoff J. S. (2000). "*Formation of highly porous biodegradable scaffolds for tissue engineering*". Electronic Journal of Biotechnology 3, 114-119. URL accessed on April 28, 2006

Mironov, V., Boland, T., Trusk, T., Forgacs, G. & Markwald, R.R., Organ printing: computer-aided jet-based 3D tissue engineering. Trends in Biotechnology 21, 157-61; 2003.

National Science Foundation (U.S.A.) (2004). "*The Emergence of Tissue Engineering as a Research Field*". URL accessed on April 28, 2006

Nerem, R.M., in Principles of Tissue Engineering. Lanza, Langer and J Vacanti (eds), 2000.

Semino, C.E., et al., Entrapment of migrating hippocampal neural cells in 3D peptide nanofiber scaffold. Tissue Engineering 10:643 (2004).

Appendix

MAJOR ENGINEERING COLLEGES IN INDIA

- Amal Jyothi College of Engineering, Kottayam, Kerala
- AWH Engineering College, Kozhikode, Kerala
- Assam Engineering College, Assam
- Amrita Insitute of Technology and Science, Coimbatore, Tamil Nadu
- Andhra University Engineering College, Vizag (AP)
- Amity School of Engineering & Technology, New Delhi
- Birla Institute of Technology of Science, Pilani 333031, Rajasthan
- Bhopal Engineering College, Madhya Pradesh
- Bharath Engineering College, Chennai, Tamil Nadu
- Basaveshwar Engineering College, Bagalkot
- Bhabha Institute of Engineering and Technology
- Birla Institute of Technology, Ranchi 835215, Bihar
- College of Engineering, Farmagadi, Goa 403401
- College of Engineering, Osmanabad
- College of Engineering and Technology, Bhubaneswar, Orissa
- C U Shah College of Engineering and Technology, Gujarat

- Delhi College of Engineering, Delhi
- D.Y. Patil College of Engineering, Maharashtra
- Engineering College in Patiala
- Datta Meghe College of Engineering, Maharashtra
- Darda Engineering College, Yavatmal, Maharashtra
- DR B R Ambedkar REC, Jalandhar, Punjab
- Hindustan College of Engineering
- Haldia Insitute of Technology, West Bengal
- Haryana Engineering College
- Indian Institute of Technology, Mumbai
- Indian Institute of Technology, Chennai
- Indian Institute of Technology, Delhi
- Indian Institute of Technology, Kanpur
- Indian Institute of Technology, Kharagpur
- JNTU College of Engineering, Hyderabad, Andhra Pradesh
- Jorhat Engineering College, Assam
- K.J. Somaiya College of Engineering
- Institute of Engineering & Technology, Kanpur, Uttar Pradesh
- Faculty of Engineering& Technology, Aligarh Muslim University, Aligarh 202002, Uttar Pradesh
- Govenment Engineering College, Maharana Pratap Nagar Bhopal 462 011, Madhya Pradesh
- G.B. Pant Engineering College, Uttar Pradesh
- Govenment Engineering College, Bhopal, Madhya Pradesh (MP).
- Government College of Engineering, Pune, Maharashtra

- LBS College of Engineering, Kasargod, Kerala
- Lala Lajpat Rai Institute of Engineering and Technology, Punjab
- Kumaraguru College of Technology, Tamil Nadu
- Karnataka Regional Engineering College, Surathkal, Srinivasnagar, Karnataka.
- Motilal Regional Engineering College, Allahabad, Uttar Pradesh
- MA College of Engineering, Kothamangalam, Kerala
- Model Engg College, Kerala
- Madan Mohan Malviya Engineering College, Gorakhpur-273010, Uttar Pradesh
- Mount Zion College of Engineering, Kadamannitta, Kerala
- M.S. Ramaiah Institute of Technology, Bangalore, Karnataka
- Malviya Regional Engineering College, Jaipur, 302107, Rajasthan
- Malnad College of Engineering, Hassan, Karnataka
- Netaji Subhas Institute of Technology, New Delhi
- Mar Baselious Christian College of Engineering & Technology, Idukki, Kerala
- Punjab Engineering College, Chandigarh, Punjab
- Periyar Maniammai College of Technology For Women, Thanjavur, Tamil Nadu
- Rajiv Gandhi College of Engineering, Research and Technology, Chandrapur
- R.V. College of Engineering , Bangalore, Karnataka.
- Regional Engineering College, Calicut-673601, Kerala
- Regional Engineering College, Tiruchirapalli, Tamil Nadu

- Regional Engineering College, Warangal, Andhra Pradesh
- Regional Engineering College, Rourkela-769008, Orissa
- Regional Engineering College, Silchur 788010, Assam
- Regional Engineering College, Kurukshetra, Haryana
- Regional Engineering College, Hamirpur, HP
- Regional Engineering College, Surat, Gujarat
- Regional Engineering College, Trichy, Kerala
- Rajagiri School of Engineering & Technology, Cochin, Kerala
- Shri Guru Gobind Singhji College of Engineering and Technology, Nanded
- Sardar Patel College of Engineering, Mumbai, Maharashtra
- Shri Ramdeobaba Kamla Nehru College-Nagpur, Maharashtra
- Sri Sant Gajanan Maharaj College of Engineering, Shegaon
- Sardar Vallabhbhai Regional College of Engineering &Technology, Surat 395007, Gujarat
- Sir M. Visvesvaraya Institute of Technology, Bangalore, Karnataka
- Sathyabama Engineering College,Chennai, Tamil Nadu
- SCMS School of Engineering and Technology, Cochin, Kerala
- TKM College of Engineering, Kollam, Kerala
- School of Planning and Architecture, Indraprastha Estate, New Delhi 110 002

- S.V. University College of Engineering, Tirupati 517502, Andhra Pradesh
- Thadomal Shahani Engineering College, Mumbai
- Terna Engineering College, Nerul, Navi Mumbai, Maharashtra
- Thapar Institute of Technology, Punjab
- University of Roorkee, Roorkee 247667, Uttaranchal (Currently IIT, Roorkee)
- University Visvesvaraya College of Engineering, Bangalore, Karnataka
- University College of Engineering, Kariyavattom, Kerala
- VeerMata Jijabai Technological Institute (VJTI) Mumbai.
- Vishveswaraya Regional College of Engineering, Nagpur 440011, Maharashtra
- Vellore Engineering College, Vellore, Tamil Nadu
- Vishwakarma Institute of Technology
- Vignana Jyothi College of Engineering & Technology
- V.E.S. Institute of Technology-Navi Mumbai, Maharashtra
- YMCA Institute of Engineering, Haryana
- Yeshwantrao Chavan College of Engineering, Maharashtra
- Zakir Hussain College of Engineering & Technology-Aligarh 202002, Uttar Pradesh

Bibliography

Adams, J. L. 1991. *Flying Buttress, Entropy, and O-Rings: The World of an Engineer*. Cambridge, MA: Harvard University Press.

Akram I. Salah (2002-04-05). Engineering an Academic Program in Software Engineering. 35th Annual Midwest Instruction and Computing Symposium.

Aldefeld B (1983) On automatic recognition of 3d structures from 2d representations. *Computer Aided Design* 15(2):59-64

Aldefeld B (1988) Variation of geometries based on a geometric reasoning method. *CAD* 20:117-126

ANSI (1982) *Dimensioning and Tolerancing*. ANSI Y14.5M, 1982

Ballard DH (1981) Generalizing the Hough Transform to Detect Arbitrary Shapes. *Pattern Recognition* 13(2):111-122

Baugh, S. G. and Roberts, R. M. (1994). Professional and organizational commitment among engineers: conflicting or complementing? *IEEE Transactions on Engineering Management*, 41: 108-8.

Bixler JP and Sanford JP (1985) A Technique for Encoding Lines and Regions in Engineering Drawings. *Pattern Recognition* 18(5):367-377

Bucciarelli, L. L. (1995). *Designing Engineers*. Cambridge: MIT Press.

Bush, V. (1946). *Endless Horizon.* Washington, D. C.: Public Affairs press.

Campbell-Allen D. and Davis, E. H. eds. (1979). *The Profession of a Civil Engineer.* Sydney: Sydney University Press.

Carter, A.G.; C. Sanger (1997). Thinking About Thinking. *The Programmers' Stone*. Alan Carter.

Chahly AT (1968) *Descriptive Geometry*. Higher School Publishing House, Moscow, 1968. (Translated from Russian by Tchernukhin AE and Botting TH).

Chai I and Dori D (1992b) Extraction of Text Boxes from Engineering Drawings. In: *Proc. SPIE/IS&T Symposium on Electronic Imaging Science and Technology, Conference on Character Recognition and Digitizer Technologies*, San Jose, CA, USA, 1992, SPIE Vol. 1661, pp 38-49

Charles R and Richard CW (1990) *The programmer¡¯s Apprentice*. ACM Press, Reading, MA

Chen Y and Langrana NA (1992) Restoration of cad database and geometric feature recognition. *Design Theory and Methodology*, DE-vol. 42:99-106

Chen Y, Langrana NA, and Das AK (1996) Perfecting Vectorized Mechanical Drawings. *Computer Vision and Image Understanding* 63(2):273-286

Chen Z and Perng D (1988) Automatic reconstruction of 3D solid objects from 2D orthographic views. *Pattern Recognition* 21:439-449

Chinese Standard (1985) *Mechanical Drawings*, (in Chinese). Mechanical Industrial Press, Beijing, China, 1985.

Chung K and Yoon J (1997) Performance Comparison of Several Feature Selection Methods Based on Node Pruning in Handwritten Character Recognition. In:

Proc. of International Conference on Document Analysis and Recognition, Ulm, Germany, August, 1997, pp 11-15

Collin S and Vaxiviere P (1991) Recognition and Use of Dimensioning in Digitized Industrial Drawings. In: *Proc. of International Conference on Document Analysis and Recognition*, Saint-Malo, 1991, pp161-169

Csink L (1989) On the Recognition of Elements Appearing In a Circuit Diagram. In: *Proc. of 2nd Hungarian AI Conference*, 1989.

Cugini U (1984) Pattern-Directed Restoration and Vectorization of Digitized Engineering Drawings. *Computers & Graphics* 8(4):337-350

Das AK and Langrana NA (1995) Recognition of dimension sets and integration with vectorized engineering drawings. In: *Proc. of Int. Conf. on Document Analysis and Recognition*, Montreal, Canada, August 1995, pp 347-340

Das AK and Langrana NA (1997) Recognition and Integration of Dimension Sets in Vectorized Engineering Drawings. *Computer Vision and Image Understanding* 68(1):90-108

David Budgen, Pearl Brereton, Barbara Kitchenham Stephen Linkman (2004-12-14). Realizing Evidence based Software Engineering.

Davies ER and Plummer APN (1981) Thinning algorithms a critique and a new methodology. Pattern Recognitio 14:53-53.

Davis, M.E., et al., Injectable Self-Assembling Peptid Nanofibers Create Intramyocardial Microenvironment for Endothelial Cells. Circulation 111:442-450 (2005

De Marco T (1978) *Structured Analysis and Syster Specification*, Yourdon Press, New York

Di Zenzo S and Morelli A (1989) A useful image representation. In: *Proc. of the Fifth International Conference on Image Analysis and Processing*, Word Scientific Publishing, Singapore, pp 170-178

Dori D (1989) A syntactic/geometric approach to recognition of dimensions in engineering machine drawings. *Computer Vision, Graphics, and Image Processing* 47(3):271-291

Dori D (1990a) Syntax Enhanced Parameter Learning for Recognition of Dimensions in Engineering Machine Drawings. *International Journal of Robotics and Automation* 5(2):59-67

Dori D (1990b) Intelligent Automatic Dimensioning of CAD Engineering Machine Drawings. *International Journal of Robotics and Automation* 5(3):124-130

Dori D (1992a) Self Structural Syntax-Directed Pattern Recognition of Dimensioning Components in Engineering Machine Drawings. In: Baird HS, Bunke H, and Yamamoto K (eds.) *Structured Document Image Analysis*, Springer-Verlag, 1992, pp 359-384

Dori D (1992b) Dimensioning analysis: Toward automatic understanding of engineering drawings. *Communications of the ACM*, 35(10):92-103, 1992.

Dori D (1994) Automated Understanding of Engineering Drawings: an Object-Oriented Analysis. *Journal of Object Oriented Programming*, Septemper 1994, 35-43

Dori D (1995a) Object-Process Analysis: Maintaining the Balance Between System Structure and Behaviour. *Journal of Logic and Computation* 5(2):227-249

Dori D (1995c) Vector-based Arc Segmentation in the Machine Drawing Understanding System Environment. *IEEE Transactions on PAMI* 17(11):959-971

Dori D (1997) Orthogonal Zig-Zag: an Algorithm for Vectorizing Engineering Drawings Compared with Hough Transform. Advances in Engineering Software 28(1):11-24

Dori D and Goodman M (1996) From Object-Process Analysis to Object-Process Design. *Annals of Software Engineering* 9:1-25

Dori D and Liu W (1997) Arc Segmentation from Complex Line Environments¡ªA Vector-Based Stepwise Recovery Algorithm. In: *Proc. of the fourth International Conference on Document Analysis and Recognition*, Ulm, Germany, August, 1997, pp 76-80

Dori D and Pnueli A (1988) The Grammar of Dimensions in Machine Drawings. *Computer Vision, Graphics and Image Processing* 42:1-18

Dori D and Tombre K (1993) Paper drawings to 3-d cad: A proposed agenda. In: *Proc. of Int. Conf. on Document Analysis and Recognition*, Tsukuba, Japan, 1993, pp 866-869

Dori D and Tombre K (1995) From engineering drawings to 3d cad models: are we ready now? *Computer Aided Design* 27(4):243-254

Dori D and Velkovitch Y (1996) Segmentation and Recognition of Dimensioning Text in Engineering Drawings. *Computer Vision and Image Understanding* (accepted on Dec. 3, 1996)

Dori D and Weiss M (1996) A Scheme for 3D Object Reconstruction from Dimensioned Orthographic Views. *Engng Applic. Artif. Intell.* 9(1):53-64

Dori D, Hubunk D, and Liu W (1997) Improving the Arc Detection Method in the Machine Drawing Understanding System. In: *Proc. of SPIE Conference on Document Analysis*, San Jose, CA, USA, Feb. 1997

Dori D, Liang Y, Dowell J, Chai I (1993) Spare Pixel Recognition of Primitives in Engineering Drawings. *Machine Vision and Applications*, 6:79-82

Dori D, Liu W, and Peleg M (1996a) How to Win a Dashed Line Detection Contest. In: Kasturi R, Tombre K (eds) *Graphics Recognition — Methods and Applications* (*Lecture Notes in Computer Science*, vol. 1072). Springer, Berlin, pp 286-300

Dori D, Phillips I, and Haralick RM (1994) Incorporating Documention and Inspection into Computer Integrated Manufacturing: An Object-Process Approach. In: Adiga S (ed.) *Applications of Object-Oriented Technology in Manufacturing*, Chapman Hall, Landon, 1994.

Dori D, Velkovitch Y, and Liu W (1996b) Object-Process Based Segmentation and Recognition of ANSI and ISO standard Dimensioning Text. In: Kasturi R, Tombre K (eds) *Graphics Recognition — Methods and Applications* (*Lecture Notes in Computer Science*, vol. 1072). Springer, Berlin, pp 212-232

Downey, G. L. and Lucena, J. C. (1995). Engineering studies. In *Handbook of Science and Technology Studies*, ed. by S. Jasanoff, G. Markle, J. Petersen, and T. Pinch, Thousand Oaks, CA: Sage; pp. 167-88.

Dunham JG (1986) Optimum uniform piecewise linear approximation of planar curves. *IEEE Transactions on PAMI* 8(1):67-75

Ejiri M, Uno T, Yoda H, Goto T, and Takeyasu K (1972) A prototype intelligent robot that assembles objects from plan drawings. *IEEE Trans. on Computers* 21(2):161-170

Fahn CS, Wang JF, and Lee JY (1988) A Topology-Based Component Extractor for Understanding Electronic Circuit Diagrams. *Computer Vision, Graphics and Image Processing* 44:119-138.

Filipski AJ and Flandrena R (1992) Automated Conversion of Engineering Drawings to CAD Form. *Proceedings of the IEEE* 80(7):1195-1209

Finch, J. K. 1961. Engineering and science: a historical review and appraisal. *Technology and Culture*, 2, 318-32.

Fletcher LA and Kasturi R (1988) A robust algorithm for text string separation from mixed text/graphics images. *IEEE Transactions on Pattern Analysis and Machine Intelligence* 10:910-918

Floreman, S. (1979). *The Civilized Engineer*. New York: St. Martin's Press.

Florman, S. C. 1976. *The Existential Pleasure of Engineering*. New York: St Martin's.

Fredrich, A. J. ed. (1989). *Sons of Martha: Civil Engineering Readings in Modern Literature*. New York: ASEE. (Many good portraits on engineers).

Fredrich, A. J. ed. 1989. *Sons of Martha: Civil Engineering Readings in Modern Literature*. New York: ASEE.

Friedel, R. (1986). Engineering in the 20th century. *Technology and Culture*, 27: 669-73.

Fukuda Y (1986) A Primary Algorithm for the Understanding of Logic Circuit Diagrams. *Pattern Recognition* 17(1):125-134

Furnas, C. C. and McCarthy, J. 1966. *The Engineer*. New York: Time Life Books.

Gao J, Tang L Liu W, and Tang Z (1995) Segmentation and Recognition of Dimension Texts in Engineering Drawings. In: *Proc. of 3rd International Conference on Document Analysis and Recognition*, Montreal, Canada, August, 1995, pp 528-531

Gerstl, J. E. and Hutton, S. P. (1966). *Engineers: The Anatomy of a Profession*. London: Tavistock Pub.

Gover, J. and Huray, P. G. (1998). The U.S. engineers shortage – How real? *Research Technology Management* 41(6): 9-14.

Groen F, Sanderson AC, and Schlag JF (1985) Symbol Recognition in Electrical Diagrams Using Probabilistic Graph Methods. *Pattern Recognition Letters*, 6:343-350

Guzman A (1968) Decomposition of a visual scene into three dimensional bodies. In: *Proc. of AFIPS Fall Joint Conf.*, San Francisco, 1968, vol. 33, pp 291-304

Hallard DH (1981) Generalizing the Hough Transform to Detect Arbitrary Shapes. *Pattern Recognition* 13(2):111-122

Haralick RM (1989) Performance assessment of near perfect machines. *Machine vision and applications* 2:1-16

Haralick RM (1992) Performance Characterization in Image Analysis¡ªThinning, a Case in Point. *Pattern Recognition Letters* 13:5-12

Haralick RM and Elliott GL (1980) Increasing tree search efficiency for constraint satisfaction problems. *Artificial Intelligence* 14:263-313

Haralick RM and Queeney D (1982) Understanding engineering drawings. *Computer Graphics and Image Processing* 20:244-258

Haralick RM and Shapiro L (1992) *Computer and Robot Vision*. Addison Wesley, Reading, MA

Haralick RM and Shapiro LG (1979) The consistent labeling problem: part 1. *IEEE Transactions on Pattern Analysis and Machine Intelligence* 1(2):173-184

Harris JF, Kittler J, Llewellyn B, and Preston G (1982) A modular system for interpreting binary pixel representation of line-structured data. In: Kittler J,

Fu KS, and Pau LF (eds) *Pattern Recognition: Theory and Applications*, D. Reidel Publisher, Norwell, Mass., 1982, pp. 311-351

He S, Abe N, and Kitahshi T (1992) Assembly planning based on assembly illustration understanding. In: *Proc. of Int. Conf. on Pattern Recognition*, Hague, Netherlands, 1992, pp 61-64

Hersh, M. (2000). The changing position of women in engineering worldwide. *IEEE Transactions of Engineering Management*, 47: 345-59.

Hilditch CJ (1969) Linear skeletons from square cupboards. *Machine Intelligence* 4:403-420

Hilliard RA, Braid IC (1978) Analysis of Dimensions and Tolerances in Computer-Aided Mechanical Design. *CAD* 10(3):161-166

Holmes, T.C., et al., Extensive neurite outgrowth and active synapse formation on self-assembling peptide scaffolds. PNAS USA 97 :6728 (2000).

Hori O and Tanigawa S (1993) Raster-to-Vector Conversion by Line Fitting Based on Contours and Skeletons. In: *Proc. of 2nd International Conference on Document Analysis and Recognition*, Tsukuba, Japan, 1993, pp 623-626

Hough PVC (1962) A method and means for recognizing complex patterns, USA Patent 3,096,654, 1962.

Hung SHY and Kasvand T (1983) Critical points on a perfectly 8- or perfectly 6-connected thin binary line. *Pattern Recognition* 16:297-284.

Hunt DJ, and Nolte LW (1988) Performance of the Hough Transform and its Relationship to Statistical Signal Detection Theory. *Computer Vision, Graphics, and Image Processing* 43:221-238

Hutton, S. and Lawrence, P. (1981). *German Engineers: The Anatomy of a Profession.* New York: Oxford University Press.

Idesawa M (1973) A System to Generate a Solid Figure from Three View. *Bulletin of JSME* 16(1):3-17

ISO (1982) *Technical Drawings*, ISO Standard Book 12, First Edition, 1982.

Iwama T, Horiuchi T, Toraichi K, Yamada H, and Yamamoto K (1994) An algorithm to restore a curved object from three orthographic views by using probabilistic relaxation matching method. In: *Proc. of IAPR Workshop on Machine Vision Applications*, Kawasaki, Japan, 1994, pp 439-442

Jacobson I, Christensson M, Jonsson P, and Overgaard GG (1992) *Object-Oriented Software Engineering*, Addison-Wesley, Reading, MA

Jaisimha MY, Haralick RM, Dori D (1993) A Methodology for the Characterization of the Performance of Thinning Algorithms. In: *Proc. of the second International Conference on Document Analysis and Recognition*, Tsukuba, Japan, pp 282-286

Jimenez J and Navalon JL (1982) Some Experiments in Image Vectorization. *IBM J. Res. Develop* 26:724-734

Joseph SH (1991) On the Extraction of Text Connected to Linework in Document Images. *Proc. of 1st International Conference on Document Analysis and Recognition*, Saint-Malo, 1991, pp 993-999

Joseph SH and Pridmore TP (1992) Knowledge-Directed Interpretation of Mechanical Engineering Drawings. *IEEE Trans. on PAMI* 14(9):928-940.

Joseph SH, Ablameyko A, and Pridmore TP (1995) Knowledge based interpretation of engineering drawings and maps. In: *Proc. of Int. Workshop on Graphics Recognition*, University Park, PA, August 1995, pp 189-199

Juster NP (1992) Modeling and Representation of Dimensions and Tolerances: A Survey. *CAD* 24(1):3-17

Kalwarski, Tara; Daphne Mosher, Janet Paskin and Donna Rosato (2006). Best Jobs in America. *MONEY Magazine*. CNN.

Kanungo T, Baird HS, and Haralick RM (1995) Estimation and validation of document degradation models. In: *Proc. of Fourth Annual Symposium on Document Analysis and Information Retrieval*, Las Vegas, Nevada, April 1995.

Kanungo T, Haralick RM, and Phillips IT (1993). Global and Local Document Degradation Models. In: *Proc. of Second International Conference on Document Analysis and Recognition*, Tsukuba, Japan, October 1993, pp 730-734

Kanungo T, Haralick RM, and Phillips IT (1994) Nonlinear local and global document degradation models. *Int. Journal of Imaging Systems and Technology* 5(4).

Karima M, Sadhal KS, and McNeil TO (1985) From paper drawings to computer-aided design. IEEE Computer Graphics and Applications 5(2):27-39.

Kasturi R and O'Gorman L (1992) Document image analysis: A bibliography. *Machine Vision Applications* 5:231-243

Kasturi R and Tombre K (eds.) (1996) *Graphics Recognition — Methods and Applications, Lecture Notes in Computer Science*, vol. 1072, Springer, May, 1996

Kasturi R, Bow ST, El-Masri W, Shah J, Gattiker JR,

Mokate UB (1990) A System for Interpretation of Line Drawings. IEEE Transactions on PAMI 12(10):978-992

Kim C, Tsuchida N, Inoue M, and Nishihara S (1992) Understanding Three-View Drawings of Mechanical Parts with Curved Shapes. In: *Proc. of IEEE International Conference on System Engineering*, Kobe (Japan), September, 1992, pp 238-241

Kimme C, Ballard DH, Sklansky J (1975) Finding circles by an array of accumulators. *Communications of the ACM* 18:120-122

Kitajima L and Yoshida M (1992) Reconstruction of CSG Solid from a Set of Orthographic Three Views. In: *Proc. of IEEE International Conference on System Engineering*, Kobe (Japan), Septemper, 1992, pp 220-224

Koen, B. V. 2003. *Discussion of the Method: Conducting the Engineer's Approach to Problem Solving*. New York: Oxford University Press.

Kong B, Phillips IT, Haralick RM, Prasad A, Kasturi R (1996) A Benchmark: Performance Evaluation of Dashed-Line Detection Algorithms. In: Kasturi R, Tombre K (eds) *Graphics Recognition — Methods and Applications* (Lecture Notes in Computer Science, vol. 1072). Springer, Berlin, pp 270-285

Lafue G (1976) Recognition of three-dimensional objects from orthographic views. ACM Computer Graphics 10(2):243-25103-108, 1976. (SIGGRAPH¡¯76 Proceedings, Philadelphia, Pennsylvania, July 14-16.)

Lai CP and Kasturi R (1991) Detection of Dashed Lines in Engineering Drawings and Maps. In: *Proc. 1st International Conference on Document Analysis and Recognition*, Saint-Malo, France, 1991, pp 507-515

Lai CP and Kasturi R (1993) Detection of Dimension Sets in Engineering Drawings. In: *Proc. 2nd International Conference on Document Analysis and Recognition*, Tsukuba, Japan, 1993, pp 606-613

Lai CP and Kasturi R (1994) Detection of dimension sets in engineering drawings. *IEEE Transactions on Pattern Analysis and Machine Intelligence* 16(8):848-855

Lam L, Lee SW, and Suen CY (1992) Thinning methodologies - A comprehensive survey. *IEEE Trans. On PAMI* 14(9):869-887

Lam L, Suen CY (1993) Evaluation of Thinning Algorithms from an OCR Viewpoint. In: *Proc. 2nd International Conference on Document Analysis and Recognition*, Tsukuba, Japan, 1993, pp 287-290

Langer, R & Vacanti JP, Tissue engineering. Science 260, 920-6; 1993 (cited more than 1000 times!)

Layton, E. T. (1971). *The Revolt of the Engineer: Social Responsibility and the American Engineering Profession.* Cleveland: Case Western University Press.

Leclerc YG and Fishler MA (1992) An optimization-based approach to the interpretation of single line drawings as 3d wire frames. *International Journal of Computer Vision* 9(2):113-136

Lee I, Dan S, Kitahashi T, and Abe N (1993) A study on a method of dividing machine parts into functional groups for technical illustrations. In: *Proc. of Int. Conf. on Document Analysis and Recognition*, Tsukuba, Japan, October 1993, pp 886-889

Lee S, Lam L, Suen CY (1991) Performance Evaluation of Skeletonization Algorithms for Document Image Processing. In: *Proc. of the first International Conference on Document Analysis and Recognition*, Saint-Malo, France, 1991, pp 260-271

Lequette R (1988) Automatic construction of curvilinear solids from wireframe views. *CAD* 20(4):171-180.

Lienhard, J. 2000. *The Engines of Our Ingenuity*. New York: Oxford University Press.

Light RA and Gossard DC (1982) Modification of Geometric Models through Variational Geometry. *CAD* 14(4):209-214

Lin X, Shimotsuji S, Minoh M, and Sakai T (1985) Efficient Diagram Understanding with Characteristic Pattern Detection. *Computer Vision, Graphics and Image Processing* 30:84-106

Liu W (1993) Knowledge-Oriented Recognition of Characters in Engineering Drawings. In: *Proc. of 3rd International Conference on CAD and Computer Graphics*, Beijing, China, 1993, pp 828-832

Liu W and Dori D (1996a) The Machine Drawing Understanding System: Incorporating Paper Drawings into CAD Environments. In: Krause F-L, Preiss K, Shpitalni M, and Shtub A (eds.) *Proc. CIEME ¨C The Computer-Integrated Extended Manufacturing Enterprise*, Agility Forum, Haifa and Herzlia, Israel, February 1996, pp 59-69

Liu W and Dori D (1997a) Recognition of Hatching Lines in Engineering Drawings. In: *Proc. 13th Israeli Symposium of AI, CV, and NN*, Tel Aviv University, Tel Aviv, Israel, February 1997. (An electronic version can be available at http://iew3.technion.ac.il:8080/~liuwy/papers/ hatchtau.ps.gz)

Liu W and Dori D (1997b) A Protocol for Performance Evaluation of Line Detection Algorithms. *Machine Vision Applications* 9(5):240-250

Liu W and Dori D (1997d) Generic Graphics Recognition of Engineering Drawing Objects. In: *Proc. of the 2nd*

IAPR Workshop on Graphics Recognition, Nancy, France, August, 1997, pp 70-77

Liu W, Dori D (1996b) Sparse Pixel Tracking: A Fast Vectorization Algorithm Applied to Engineering Drawings. In: *Proc. of the 13th International Conference on Pattern Recognition*, Vienna, Austria, Vol. III (Robotics and Applications), pp 808-812

Liu W, Dori D (1996c) Automated CAD Conversion with the Machine Drawing Understanding System. In: *Proc. of the International Workshop on Document Analysis Systems*, Malvern, PA, USA, pp 241-259

Liu W, Dori D, Tang L, Tang Z (1995b) Object Recognition in Engineering Drawings Using Planar Indexing. In: *Proc. of the 1st IAPR Workshop on Graphics Recognition*, Penn. State Univ., USA, 1995, pp 53-61

Liu W, Tang L, and Tang Z (1995a) Processing and Understanding of Engineering Drawings: Toward Facilitating the Reuse of Engineering Drawings. In: *Proc. of 5th International Conference on Computer Aided Product and Engineering*, Beijing, China, May 1995, pp 151-160

Love, A. and Childers, J. S. eds. (1965). *Listen to Leaders in Engineering*. Atlanta, GA: Tupper & Love.

Love, A. and Childers, J. S. eds. 1965. *Listen to Leaders in Engineering*. Atlanta, GA: Tupper & Love.

Lysak DB and Kasturi R (1990) Interpretation of line drawings with multiple views. In: *Proc. of Int. Conf. on Pattern Recognition*, Atlantic City, 1990

Lysak DB and Kasturi R (1991) Interpretation of engineering drawings of polyhedral and nonppolyhedral objects. In: *Proc. of Int. Conf. on Document Analysis and Recognition*, Saint-Malo, France, 1991, pp 79-87

Lysak DB and Kasturi R (1995) View labeling for automated interpretation of engineering drawings. *Pattern Recognition* 28(3):393-407

Ma, Peter X.: Scaffolds for tissue fabrication - *Materials Today*, May 2004, 30-40.

MacArthur, B. D. & Oreffo, R.O.C. (2005). "*Bridging the gap*". Nature 433, 19.

Madej D. and Sokolowski A (1993) Towards automatic evaluation of drawing analysis performance: A statistical model of cadastral map. In: *Proc. of Int. Conf. on Document Analysis and Recognition*, Tsukuba, Japan, October 1993, pp 890-893

Markowsky G and Wesley MA (1980) Fleshing out wireframes. *IBM Journal of Research and Development* 24(5):582-597

Marston RE and Kuo MH (1994) Reconstruction of 3D objects from three orthographic projections using a decision-chaining method. In: *Proc. of IAPR Workshop on Machine Vision Applications*, Kawasaki, Japan, December 1994, pp 423-426

Marston RE and Kuo MH (1995a) Efficiently identifying the faces of a wireframe reconstructed from 3-view engineering drawings. In: *Proc. of Symposium on Document Analysis and Information Retrieval*, Las Vegas, Nevada, April 1995, pp 189-199

Marston RE and Kuo MH (1995b) Interpretation of 3-view engineering drawings as central quadric surface mechanical parts. In: *Proc. of Int. Conf. on Document Analysis and Recognition*, Montreal, Canada, August 1995, pp 343-346

Marti E, Regincos J, Lopez-Krahe J, and Villanueva JJ (1991) A System for Interpretation of Hand Line Drawings as Three-Dimensional Scene for CAD Input.

In: *Proc. of Int. Conf. on Document Analysis and Recognition*, Saint-Malo, France, 1991, pp 472-480

Mascone, C. F. 1999. Engineering the next millennium. *Chemical Engineering Progress*, 95(10): 102-13.

McConnell, Steve *(August 2003). Professional Software Development: Shorter Schedules, Better Projects, Superior Products, Enhanced Careers. Boston, MA: Addison-Wesley.*

Meehan, R, L, (1981). *Getting Sued and Other Tales of the Engineering Life*. Cambridge: MIT Press.

Meikins, P. and Smith, C. (1996). *Engineering Labor: Technical Workers in Comparative Perspective*. London: Verso.

Mikos, A. G. & Temenoff J. S. (2000). *"Formation of highly porous biodegradable scaffolds for tissue engineering"*. Electronic Journal of Biotechnology 3, 114-119. URL accessed on April 28, 2006

Min W, Tang Z, and Tang L (1993) Recognition of Dimensions in Engineering Drawings Based on Arrowhead-match. In: *Proc. of 2nd International Conference on Document Analysis and Recognition*, Tsukuba, Japan, 1993, pp 373-376

Min W, Tang Z, and Tang L (1994) Using Web Grammar to Recognize Dimensions in Engineering Drawings. *Pattern Recognition* 26(9):

Mironov, V., Boland, T., Trusk, T., Forgacs, G. & Markwald, R.R., Organ printing: computer-aided jet-based 3D tissue engineering. Trends in Biotechnology 21, 157-61; 2003.

Monagan G and Roosli M (1993) Appropriate Base Representation Using a Run Graph. In: *Proc. of 2nd International Conference on Document Analysis and Recognition*, Tsukuba, Japan, 1993, pp 623-626

Montanari U (1969) Continuous skeletons from digitized images. *Journal of the ACM* 16:534-549.

Montanari U (1970) A note on the minimal length polygonal approximation to a digitized contour. *Communications of the ACM* 13(1):41-47.

Naccache NJ and Shinghal R (1984a) An investigation into the skeletonization approach of Hilditch. *Pattern Recognition* 17:279-284.

Naccache NJ and Shinghal R (1984b) SPTA: a proposed algorithm for thinning binary patterns. *IEEE Trans. on SMC* 14:409-418

Nagasamy V and Langrana N (1990) Engineering Drawing Processing and Vectorization System. *Computer Vision, Graphics and Image Processing* 49(3):379-397

Nagasamy V and Langrana N (1990) Reconstruction of three-dimensional objects using a knowledge-based environment. *Engineering with Computers* 7:23-35

Nagendra IV and Gujar UG (1988) 3-d objects from 2-d orthographic views¡ªa survey. *Computers and Graphics* 12(1):111-114

Nalwa VS (1993) A Guided Tour of Computer Vision. Addison-Wesley, New York.

National Science Foundation (U.S.A.) (2004). *"The Emergence of Tissue Engineering as a Research Field"*. URL accessed on April 28, 2006

Nerem, R.M., in Principles of Tissue Engineering. Lanza, Langer and J Vacanti (eds), 2000.

Pao D, Li HF, and Jayakumar R (1991) Graphic Feature Extraction for Automatic Conversion of Engineering Line Drawings. In: *Proc. 1st International Conference on Document Analysis and Recognition*, Saint-Malo, France, 1991, pp 533-541

Parnas, David L. *(1998)*. "Software Engineering Programmes are not Computer Science Programmes". *Annals of Software Engineering 6: 19–37.*, p. 19.

Parnas, David L. *(1998)*. "Software Engineering Programmes are not Computer Science Programmes". *Annals of Software Engineering 6: 19–37.*, p. 20.

Pasternak B and Neumann B (1993) Adaptable drawing interpretation using opbject-oriented and constraint-based graphic specification. In: *Proc. of Int. Conf. on Document Analysis and Recognition*, Tsukuba, Japan, October 1993, pp 359-364

Pecht, Michael (1995). Product Reliability, Maintainability, and Supportability Handbook. CRC Press.

Pehrson, Ronald J. (January 1996). "Software Development for the Boeing 777". CrossTalk: The Journal of Defense Software Engineering.,

Petroski, H. 1982. *To Engineer is Human: the Role of Failure in Successful Design*. New York: St. Martin's Press.

Petroski, H. 1996. *Design Paradigms: Case Histories of Errors and Judgment in Engineering*. New York: Cambridge University Press.

Petroski, H. 1996. *Invention by Design*. Cambridge, MA: Harvard University Press.

Pfaltz JL and Rosenfeld A (1967) Computer representation of planar regions by their skeletons. *Communications of the ACM* 10:119-125

Preiss K (1981) Algorithms for automatic conversion of a 3-view drawing of a plane-faced part to the 3-D representation. *Computers in Industry* 2:133-139

Preiss K (1984) Constructing the solid representation from engineering projections. *Computers and Graphics* 8(4):381-389

Requicha AAG (1980) Representation for rigid solids: theory, methods, and systems. *ACM Computing Surveys* 12(4):437-464

Requicha AAG and Voelcker HB (1983) Solid modeling: Current status and research directions. *IEEE Computer Graphics and Applications* 2(2):9-24

Rumbaugh J, Blaha M, Premerlani W, Eddy F, and Lorensen W (1991) *Object Oriented Modeling and Design*. Prentice Hall International Editions, Englewood Cliffs, NJ, 1991.

Semino, C.E., et al., Entrapment of migrating hippocampal neural cells in 3D peptide nanofiber scaffold. Tissue Engineering 10:643 (2004).

Simon, H. (1996). *The Sciences of the Artificial*, 3rd ed. Cambridge: MIT Press.

Smiles, S. (1874). *Lives of the Engineers*. London: John Murray.

The 1968 NATO Software Engineering Conference

U.S Department of Labor and Statistics *The 2000 Standard Occupational Classification (SOC) System: 15-0000 Computer and Mathematical Occupations*

Vaughan, D. (1990). The image of the engineer in the popular imagination, 1880-1980. *Bulletin of Science, Technology, and Society*, 10: 301-4.

Vincenti, W. G. (1990). *What Engineers Know and How They Know It: Analytical Studies from Aeronautical History*. Baltimore, MD: Johns Hopkins Press.

Wallich, P. (1984). The engineer's job: it moves toward abstraction. *IEEE Spectrum*, 21(6): 32-7.

Williams, R. (2000). "All that is solid melts into air:" Historians of technology in the information revolution. *Technology and Culture* 41: 641-68.

Index

D

E

F

R

S

T

U

V

■■■